D0315814

Construction technology 1:
HOUSE CONSTRUCTION

THIRD EDITION

Mike Riley

Director of the School of the Built Environment,
Liverpool John Moores University, UK

and

Alison Cotgrave

Deputy Director of the School of the Built Environment,
Liverpool John Moores University, UK

palgrave
macmillan

This edition first published 2013 by
PALGRAVE MACMILLAN

Palgrave Macmillan in the UK is an imprint of Macmillan Publishers Limited,
registered in England, company number 785998, of Houndmills, Basingstoke,
Hampshire RG21 6XS.

Palgrave Macmillan in the US is a division of St Martin's Press LLC,
175 Fifth Avenue, New York, NY 10010.

Palgrave Macmillan is the global academic imprint of the above companies
and has companies and representatives throughout the world.

Palgrave® and Macmillan® are registered trademarks in the United States,
the United Kingdom, Europe and other countries.

ISBN 978–1–137–03017–7   paperback

This book is printed on paper suitable for recycling and made from fully
managed and sustained forest sources. Logging, pulping and manufacturing
processes are expected to conform to the environmental regulations of the
country of origin.

A catalogue record for this book is available from the British Library.

A catalog record for this book is available from the Library of Congress.

Contents

Acknowledgements

The authors would like to express their thanks to the following people for their support and contribution to the book:

Julie, Steve and Sam:
>   for their support and understanding during the writing of this text

Paul Hodgkinson and Dave Tinker:
>   for their time and effort in creating the illustrations contained within the book

Robin Hughes:
>   for his advice and commentary on the preliminary drafts

Redrow Homes and W. A. Browne Ltd:
>   for their kind permission to use examples of their buildings for illustrating the case studies and for granting unlimited access to their construction sites

Steel Construction Institute:
>   for kind permission to reproduce certain images from their guidance documentation

'Polysteel':
>   for permission to reproduce images in connection with insulated concrete formwork construction

Steve Allin, Sustainable Hemp Consultant, Rusheens Kenmare:
>   for permission to use photographs of hemp construction shown in Figures 13.7 and 13.8

Denova Design, Liverpool:
>   for photographs of sustainable housing

Peter Williams:
>   for his contribution and help with the section relating to Health & Safety

Linda Howes
>   for assisting with the written amendments

Sergejs Katkovskis and Fotolia.com
>   for kind permission to use the photo in chapter 5, on page 134

Preface

There are many texts available in the area of building technology, each with their own merits. Each new book in the field attempts to provide something extra, something different to set it apart from the rest. The result is an ever-expanding quantum of information for the student to cope with. Whilst such texts are invaluable as reference sources, they are often difficult to use as learning vehicles. This text seeks to provide a truly different approach to the subject of construction technology associated with houses. Rather than being a reference source (although it may be used as such), the book provides a learning vehicle for students of construction and property-related subjects. It has been updated to incorporate changes in Building Regulations and other legislation, and this third edition includes recent innovations in residential construction, particularly in the area of sustainable construction.

A genuine learning text

The text is structured so that it provides a logical progression and development of knowledge from first principles to more advanced concepts relating to the technology of house construction. The content is aimed at students wishing to gain an understanding of the subject matter without the need to consult several different and costly volumes. Unlike reference texts that are used for selectively accessing specific items of information, this book is intended to be read as a continuous learning support vehicle. Whilst the student can access specific areas for reference, the major benefit can be gained by reading the book from the start, progressing through the various chapters to gain a holistic appreciation of the various aspects of house construction.

Key learning features

- The learning process is supported by several key features that make this text different from its competitors. These are embedded at strategic positions to enhance the learning process.
- Case studies include photographs and commentary on specific aspects of the technology of house construction. Thus students can visualise details and components in a real situation.
- Reflective summaries are included to encourage the reader to reflect on the subject matter and to assist in reinforcing the knowledge gained.

- Review tasks aid in allowing the reader to consider different aspects of the subject at key points in the text.
- Comparative studies allow the reader to quickly compare and contrast the features of different details or design solutions and are set out in a simple-to-understand tabular format.
- The Info points incorporated into each chapter identify key texts or sources of information to support the reader's potential needs for extra information on particular topics.
- In addition, margin notes are used to expand on certain details discussed in the main text without causing readers to divert their attention from the core subject matter.

Website

A website supporting this book and designed to enhance the learning process can be found at www.palgrave.com/engineering/riley1. This is separated into a lecturer's zone and a student zone.

- The lecturer's zone has series of 'studios' that the lecturer can use with students to support the learning process. The studios are designed to build upon the information contained within the book to provide a challenging and interesting series of case-based learning situations. The studios are set around Case study material and they encourage the students to access and interpret information from a variety of defined sources such as British Standards, BRE Digests and other published material. Each of the studios is structured to create a scenario within which students are required to consider a range of options for the solution of design and construction problems based on the knowledge developed by using the book, but also requiring wider, directed research.

  The studios may be used as learning vehicles, tutorial tasks and assessed submissions if delivered appropriately. In addition to the studio tasks the lecturer's zone incorporates detailed descriptions relating to the Case study materials together with additional photographs and visual materials suitable for use in formal lectures and tutorials to follow the structure and format of the book.
- The student zone contains outline answers to the Review tasks, plus further photographs from the Case studies in the book, with accompanying commentary.

MIKE RILEY
ALISON COTGRAVE

Introduction to house construction

1

Functions of buildings

AIMS

After studying this chapter you should be able to:

- Appreciate the main physical functions of buildings
- Describe the factors that must be considered in creating an acceptable living environment
- Discuss links between these factors and the design of modern dwellings
- Recognise the sources and nature of loads applied to building elements and the ways in which they affect those elements
- Appreciate the influence of the choice of materials and the selection of design features on building performance

This chapter contains the following sections:

1.1 Physical and environmental functions of buildings
1.2 Forces exerted on and by buildings
1.3 Structural behaviour of elements

INFO POINT

- Building Regulations Approved Document A, Structure (2004 including 2010 amendments)
- BS 648: Schedule of weights of building materials (1964 – withdrawn)
- BS 5250: Code of practice for control of condensation in buildings (2011)
- BS ISO 6243: Climatic data for building design. Proposed system of symbols (refers to structural design of buildings) (1997)
- BS 6399 Part 1: Loadings for buildings. Code of practice for dead and imposed loads (1984)
- BS EN ISO 7730: Moderate thermal environments. Determination of the PMV and PPD indices and specification of the conditions for thermal comfort (1995)

1.1   Physical and environmental functions of buildings

Introduction

- After studying this section you should be aware of the nature of buildings as environmental enclosures.
- You should appreciate the nature of the building user's need to moderate the environment and understand how built form has evolved to allow this to be achieved.
- You should have an awareness of the link between environmental needs and the form of the building fabric.
- You should be able to recognise the key features of building form that affect the internal environment.
- You should understand the nature of physical forces exerted on and by buildings and you should be familiar with the terminology associated with this aspect of building performance.
- You should have comprehension of the implications of the need to satisfy these requirements upon building design.
- Given a variety of scenarios you should be able to recognise the key features of the building structure and fabric and should be able to relate these to the building's physical and environmental performance.

Included within this section are the following areas:

- The building as an environmental envelope
- Performance requirements of building fabric

The building as an environmental envelope

Overview

The ways in which the internal environments of buildings are controlled have become very sophisticated as the needs of occupiers have evolved. The degree to which we are able to moderate the internal environmental conditions using the building enclosure and building services is great. However, it is easy to take for granted some of the features of buildings that affect the internal environment and to overlook the basis of the evolution of these features. Dwellings are generally designed to be aesthetically pleasing. Many of the details that we associate with building style and aesthetics have their origins in the need to satisfy functional needs. As buildings have developed, the role of building services to control heat, light and ventilation has become more significant. It is easy to forget that these services rely on the existence of an appropriate building envelope in order to achieve the required level of performance. The dwelling as we now know it has its origins in the simplest form of building enclosure, created by people to protect them from the extremes of the environment. The factors that led people to develop such enclosures in historic times are still evident today, and the function of dwellings, although now much more sophisticated, is still essentially the same as it was then. One of the primary functions of the building fabric is to create an environmental envelope.

Buildings and the control of the internal environment

Historically people have sought to modify and control the environment in which they live. In prehistoric times caves and other naturally occurring forms of shelter were used as primitive dwellings, providing protection from the external environment. As civilisation has developed, so the nature of people's shelter has become more refined and complex, developing from caves and natural forms of shelter to simple artificial enclosures, such as those used throughout history by nomadic peoples worldwide. The ways in which the structures created by humans developed have depended upon the nature of the climate in specific locations and the form of building materials available locally. This resulted in the development of vernacular forms of building, based on the use of readily available local materials. As a consequence, identifiable styles of buildings developed in different areas, each adapting the form of the building to satisfy functional requirements with available materials and technologies. The ability to transport building materials over relatively large distances is a recent development. In Britain, for example, this was limited prior to the Industrial Revolution by the lack of effective transport networks. The advent of canals and rail links allowed materials to be transported over relatively large distances. Hence the extent of vernacular architecture has reduced, with materials from a wide variety of locations being incorporated into more modern buildings to satisfy functional requirements in the most efficient and cost-effective way possible.

Examples of vernacular architecture are found in the UK and throughout the world. In areas such as the Middle Eastern desert regions, where **diurnal temperatures** vary considerably, being very hot during the day and cool at night, buildings of massive construction are common. Such buildings are referred to as 'thermally heavy' structures. The intense heat of the day is partly reflected by the use of white surface finishes, and that which is not reflected is absorbed by the building fabric rather than being transmitted into the occupied space. As a result of the slow thermal reaction of the building this stored heat is released at the times of day when the external temperatures may be very low, acting as a form of storage heater. The effects of direct solar gain are reduced by the use of a limited number of small window openings.

In areas where the climate is consistently warm and humid, such as in South East Asia, a very different approach to building design is required. In such situations, rare breezes may be the only cooling medium that can remove the oppressive heat and humidity of the internal environment. Since this cooling and dehumidifying effect takes place in a short period, the building must be able to react quickly to maximise any potential benefit. Hence, a 'thermally light' structure is essential to transmit external changes to the interior with minimal delay. The nature of buildings in such areas reflects these requirements, with lightweight building fabric and many large openings to allow cooling breezes to pass through the building (Figure 1.1).

During the course of a typical day the air temperature will vary as a result of climatic conditions. The variation of **diurnal temperatures** during a single day is referred to as the *diurnal range*.

Figure 1.1
Thermal response of
building fabric.

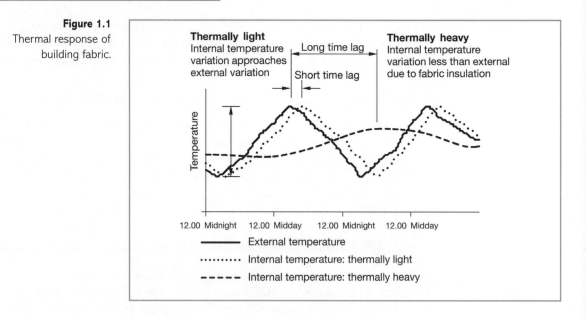

There is a great difference in the properties of thermally light buildings and thermally heavy buildings. The fabric of the thermally light building is generally light in weight, with little capacity to absorb and store heat. Thus these buildings display fast response to external temperature changes. Thermally heavy buildings, in contrast, tend to be massive in their construction form, with dense walls that absorb heat readily. This construction form insulates the interior from external changes as a result of slow reaction times. The selected construction form must be matched to climate, user needs and the building services that are present to actively modify the internal environment.

The use of protective structures or enclosures is not the only method utilised in the moderation of people's environments. Since fire was first discovered and used by primitive people to provide light and heat, the use of energy to aid in environmental moderation has been fundamental. Although the use of built enclosures can moderate the internal environment and reduce the effects of extremes in the external climate, the active control and modification of the internal environment require the input of energy. The use of buildings to house people, equipment and processes of differing types, exerting differing demands in terms of internal environment, has resulted in the development of buildings and associated services capable of controlling the internal conditions within desired parameters with great accuracy.

The nature of people's perceptions of comfort within buildings has also developed. The simple exclusion of rain and protection from extreme cold or heat are no longer sufficient to meet human needs. The provision of an acceptable internal condition relies on a number of factors, which include:

- *Thermal insulation and temperature control*: The fabric of a modern dwelling must ensure that the levels of heat transfer between the interior and the external environment are within acceptable limits. In some cases this is aimed at minimising heat loss during cold periods; in others the aim is to minimise heat gain. In both of these situations the fabric must possess good levels of insulation in order to fulfil its functional requirement.

- *Acoustic insulation*: In most situations it is desirable to shield the internal environment within a dwelling from the noise of the external surroundings. In addition, the need to maintain levels of privacy demands that the external envelope of the building be capable of insulating against the passage of noise to a reasonable level.

- *Provision of light*: The interior of a dwelling must be provided with sufficient levels of natural or artificial light to allow the users to undertake their daily activities without hindrance. In addition, the levels of lighting maintained affect the perceptions of the comfort of the internal environment. Areas that are not provided with natural lighting would not be considered as being fit for human habitation.

- *Control of humidity and ventilation*: In any building environment there is an acceptable range of relative humidity within which most people will feel comfortable. If the environment is subject to humidity levels above or below this range the occupants will feel discomfort; hence the humidity levels must be controlled. This is normally achieved in dwellings by the provision of appropriate levels of ventilation. The daily activities undertaken within a dwelling produce significant amounts of water vapour from cooking, washing and so on. Ventilating the interior to allow this vapour to migrate to the exterior is an effective method of humidity control. The provision of opening window areas facilitates ventilation and air movement within the dwelling, providing a constantly changing air supply to the interior.

- *Exclusion of contaminants*: One of the functions fulfilled by the provision of appropriate levels of ventilation is the effective removal or exclusion of contaminants. Smoke, odours and other contaminants that are present in the air will be removed as a result of providing appropriate levels of air movement and, more importantly, air change through ventilation.

The extent to which each of these plays a part in the creation of an acceptable environment varies from situation to situation. As the construction industry seeks to develop a more sustainable approach to its activities there is a move towards more natural mechanisms for environmental control. Many new buildings are designed to maximise natural light and ventilation, with the aim of reducing energy costs. Similarly, the benefits of solar gain may be maximised in the design of dwellings with differing sizes of window on different elevations depending on orientation. Large windows on southerly elevations maximise solar gain, while small openings on northerly elevations minimise heat loss. In the Northern hemisphere the sun shines from the south. Hence most solar gain is on south-facing elevations. This affects the design of buildings such that larger windows tend to be placed to maximise heat gain, with small windows on north-facing walls to reduce heat loss. Factors such as this will play an increasing part in the design of buildings as we move towards a more sustainable approach to the built environment.

Figure 1.2
The building as an
environmental modifier.

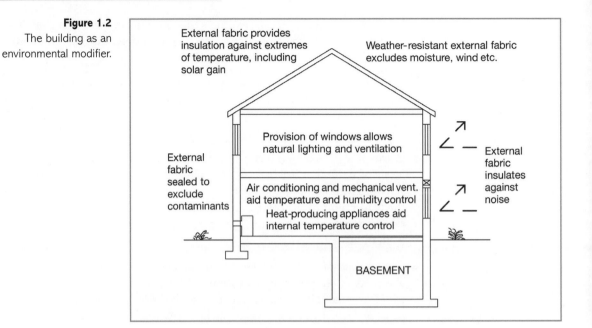

Figure 1.2 illustrates these requirements and the ways in which they are met in modern construction forms.

The construction of buildings using materials that were readily available locally was common in Britain prior to the Industrial Revolution. After this the ability to transport materials via the canal and railway systems resulted in more widespread use of some materials. The adoption of slate for roof coverings is a good example of this. In the UK the local availability of slate is limited to quite restricted areas, such as North Wales and the Lake District. However, the use of slate as a roofing material

Figure 1.3
Vernacular forms of
architecture.

These dwellings are located in Thailand. The lightweight construction form is created by the use of locally available materials. The large openings allow air to pass through the building to cool the interior quickly.

The term **vernacular building** style is used to relate to buildings that are constructed using local materials and details that are specific to the region in which they are built.

was widespread. In some countries the use of local materials and techniques is still commonplace and the wide variety of **vernacular building** styles reflects this. As previously noted, the form of these buildings is also affected by the nature of the local climate and the need to moderate the internal environment. Figure 1.3 shows an example of a regional vernacular style.

The sustainability agenda is growing in acceptance and prominence worldwide. As a consequence of this, the impact of construction and urban development is under scrutiny. One of the results of this scrutiny is an increased awareness of a building's environmental footprint. This is driving a return to the use of locally sourced materials and materials that have been sourced in an environmentally responsible manner.

REVIEW TASKS

■ What performance differences do we expect from *thermally light* buildings compared with *thermally heavy* buildings?

■ Undertake an Internet search using the terms 'vernacular building' and 'modern house construction'. Identify the key differences in the images that you locate. How do these relate to the performance requirements that we have previously identified?

■ Visit the companion website at www.palgrave.com/engineering/riley1 to view sample outline answers to the review tasks.

Performance requirements of building fabric

Overview

The requirement to provide an acceptable internal environment is merely one of the performance requirements of modern buildings. The level of performance of buildings depends upon several factors and the emphasis placed upon individual performance requirements varies from situation to situation. Statutes and guidelines, such as the Building Regulations, set out minimum standards. These standards must always be satisfied irrespective of perceptions of the performance of the building fabric. The increasing role of the building as an asset has also affected the ways in which buildings have been designed to maximise the long-term value and to minimise the maintenance costs of the structure and fabric. House construction is linked strongly to 'marketability' in some countries.

The performance requirements of buildings include:

■ Structural stability
■ Durability
■ Thermal insulation
■ Exclusion of moisture and protection from weather
■ Acoustic insulation

- Flexibility
- Aesthetics
- Buildability
- Sustainability.

Each of these performance requirements is important, although in certain instances some aspects of performance may take on more significance than others. A good example of this is the issue of durability. The level of durability required is linked to the required lifespan of the building. In some cases the intended functional life of the building may be short. In these situations the level of durability required clearly takes on less importance than in a situation where the lifespan is intended to be longer. Notwithstanding this, the relative importance of the various requirements is generally considered to be equal, and each of the requirements must be addressed to some extent. It is therefore worthwhile considering each in turn.

Structural stability

In order to satisfactorily fulfil the functions required of it, a building must be able to withstand the loads imposed upon it without suffering deformation or collapse. This necessitates the effective resistance of loads or their transfer through the structure to the ground. In dwellings of traditional structure the mechanisms for dealing with these loads are rather different from those of a timber frame or modular building. The principles of each are dealt with later in this book and will not be explored in detail here. However, it is important to note that whatever the form of the building structure, the loads must be dealt with effectively. Generally this is achieved either by transferring them to some intermediate supporting element or to the supporting strata. Individual elements or components of the structure must also possess sufficient strength to cope with the forces that are established within them as a result of the various forms of load that exist within buildings.

Durability

The long-term performance of the structure and fabric demands that the component parts of the building are able to withstand without deterioration the vagaries and hostilities of the environment in which they are placed. The ability of the parts of the building to maintain their integrity and functional ability for the required period of time is fundamental to the ability of the building to perform in the long term. This factor is particularly affected by the occurrence of fires in buildings. The issue of durability affects the design of the building fabric and the selection of building materials and components. Great care must be taken in the specification of materials and components, as well as in the detailed design of the building, if premature failure is to be avoided. Durability is generally considered to be a variable benchmark of building performance, in that it is linked to the intended design life of the building rather than being an absolute measure of performance.

Thermal insulation

The needs to maintain internal conditions within fixed parameters and to conserve energy dictate that the external fabric of a given building provides an acceptable standard of resistance to the passage of heat. The level of thermal insulation that is desirable in an individual instance is, of course, dependent upon the use of the building, its location and so on. As the costs of energy increase and the awareness of environmental issues becomes more widespread, so the issue of energy consumption increases in importance. The Building Regulations set down minimum requirements for thermal performance of building enclosures. These requirements will undoubtedly increase in the future, and the design of the building fabric will evolve accordingly.

Exclusion of moisture and protection from weather

The passage of moisture from the exterior, whether in the form of ground water rising through capillary action, precipitation or other possible sources, should be resisted by the building envelope. The ingress of moisture to the building interior can have several undesirable effects. These include the decay of timber elements, deterioration of surface finishes and decorations and risks to the health of the occupants, in addition to effects upon certain processes carried out in the building. Hence, details must be incorporated into the design of the building structure and fabric to resist the passage of moisture to the interior of the building from all undesirable sources. The exclusion of wind and water is essential to the satisfactory performance of any building fabric. In addition, the associated issue of exclusion of contaminants is increasingly recognised. One example of this is the potentially **deleterious** effect of radioactive radon gas upon the occupants of buildings. In areas where this is likely, specific design details must be introduced to minimise the potential risk associated with ingress of the gas.

Deleterious is a term used to mean 'harmful' to mind or body. In the context of buildings it is often used with reference to hazardous materials.

Acoustic insulation

The passage of sound from the exterior to the interior, or between interior spaces, should be considered in building construction. The level of sound transmission that is acceptable in a building will vary considerably, depending upon the nature of the use of the building and its position. In the case of dwellings this is of particular concern where individual dwelling units are adjacent or contained within the same building enclosure. In fact, this is the case for the vast majority of dwellings in the UK. Free-standing or detached dwellings are far less common than semi-detached and terraced houses or flats. The transmission of noise between linked dwelling units can be a major problem, and all new buildings take this into account. There is also an issue associated with the transfer of sound between the exterior and the interior of the building. This can be of particular concern in locations where there is likely to be significant noise, as in the case of dwellings close to airports, for example.

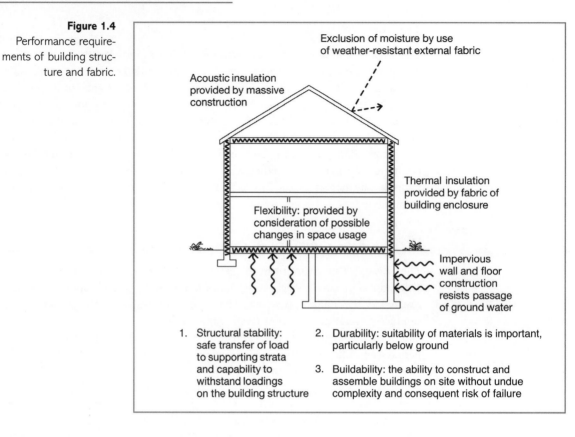

Figure 1.4
Performance requirements of building structure and fabric.

Exclusion of moisture by use of weather-resistant external fabric

Acoustic insulation provided by massive construction

Thermal insulation provided by fabric of building enclosure

Flexibility: provided by consideration of possible changes in space usage

Impervious wall and floor construction resists passage of ground water

1. Structural stability: safe transfer of load to supporting strata and capability to withstand loadings on the building structure

2. Durability: suitability of materials is important, particularly below ground

3. Buildability: the ability to construct and assemble buildings on site without undue complexity and consequent risk of failure

Flexibility

In industrial and commercial buildings in particular, the ability of the building to cope with and respond to changing user needs has become very important. Hence the level of required future flexibility must be taken into account in the initial design of the building. This is reflected, for example, in the trend to create buildings with large open spaces that may be subdivided by the use of partitions which may be readily removed and relocated. In reality, although the structural form of dwellings may well lend itself to flexibility in layout, this is rarely exploited by occupiers.

Aesthetics

The question of building aesthetics is subjective; however, it should be noted that in some situations the importance of the building's aesthetics is minimal, whereas in others it is of course highly important. For example, the appearance of a unit on an industrial estate is far less important than that of a city centre municipal building. The extent to which aesthetics are pursued will have an inevitable effect on the cost of the building. In the UK and other places there has been a tendency for house building to follow traditional design, as this is most readily marketable.

Buildability

In recent years developers and constructors have made greater progress in reducing the number of defects in buildings. A link has been recognised between the complexity of details and the probability of failure. Hence great efforts are now made to ensure that buildings and their component details are physically 'buildable' in true site conditions.

This summary is not a definitive list of the performance requirements of all building components in all situations; however, it is indicative of the factors which affect the design and performance of buildings and their component parts (Figure 1.4).

Sustainability

There is now a widespread acceptance of the need to undertake all construction activity, including the construction of dwellings, in a sustainable way. This is supported by legislative controls and drivers such as the Building Regulations and a raft of specific environmental legislation. In the UK the introduction of energy ratings for new homes and the requirement to include energy information in the sales details of existing homes have ensured that sustainability is an issue that is recognised not just by constructors but also by dwelling owners and occupiers.

REVIEW TASKS

- List *five* performance requirements that you hope will be supplied by a dwelling, and rank them in order of importance from a resident's point of view.

- Identify a recently constructed building and another that was built at least 30 years ago. Consider the ways in which they each attempt to meet the performance requirements defined above.

- Visit the companion website at www.palgrave.com/engineering/riley1 to view sample outline answers to the review tasks.

1.2 | Forces exerted on and by buildings

Introduction

- After studying this section you should be aware of the differing forces that act upon the structure of buildings.
- You should have developed knowledge of the origins of these forces and the nature in which they act upon the structural elements of the building.
- You should have an intuitive appreciation of the relative magnitude of various forces.

- You should be able to distinguish between the forces exerted on the building and the forces exerted by it.
- You should understand the terminology associated with the forces active within buildings, and, given a variety of scenarios, be able to recognise the different types of forces and appreciate their implications.

Loading forms

Overview

Forces acting upon the structural elements of buildings derive from a variety of sources and act in many different ways. However, there are a number of basic principles of structural behaviour, and these can be applied in considering the application and effect of the various forces that act upon buildings. The ways in which the structure and fabric of a building behave will depend upon their ability to cope with a range of inherent and applied loads or 'dead' and 'live' loads. If the building is able to withstand the loading imposed upon it, it will remain static; in such a state it is considered to be stable. Any force acting upon a building is considered as a loading, whether it arises as a result of external factors, such as the action of wind on the building, or from the use of the building, such as the positioning of furniture, equipment or people. We must also recognise that the building itself is a source of loading simply as a result of the effect of the self-weight of the structure.

In order to withstand these forces two basic structural properties must be provided by the building. First, the component parts of the building must possess adequate strength to carry the applied loads. Second, the applied forces must be balanced in order to resist the tendency for the building to move. Thus the building must remain in equilibrium. In order to understand how these factors are achieved we must first examine the nature of the loads that act upon buildings.

The nature of loads

The forces, or loads, applied to buildings can be considered under two generic classifications: *dead loads* and *live loads*. Dead loads would normally include the self-weight of the structure, including floors, walls, roofs, finishes, services and so on. Live loads would include loads applied to the building in use, such as the weight of people, furniture, machinery and wind loads. Such loads are normally considered as acting positively on the building. However, in the case of wind loads (Figure 1.5) suction zones may be created (that is, negative forces); this effect is often illustrated by the action of roofs being lifted from buildings in high wind conditions. Hence buildings must be designed to cope with forces acting in a variety of ways.

The ability of the materials used in the construction of buildings to withstand these loads is termed 'strength'. In considering whether a building has sufficient strength, the different types of load must be considered. Their direction is also important: they may be oblique (at an angle) or axial (along the axis of an element).

PART 1

Figure 1.5
Effects of wind loads on a
building.

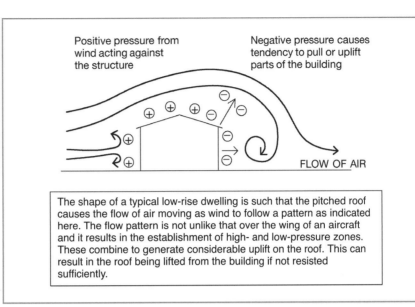

Positive pressure from
wind acting against
the structure

Negative pressure causes
tendency to pull or uplift
parts of the building

FLOW OF AIR

The shape of a typical low-rise dwelling is such that the pitched roof
causes the flow of air moving as wind to follow a pattern as indicated
here. The flow pattern is not unlike that over the wing of an aircraft
and it results in the establishment of high- and low-pressure zones.
These combine to generate considerable uplift on the roof. This can
result in the roof being lifted from the building if not resisted
sufficiently.

Stress

When subjected to forces, all structural elements tend to deform. This deformation
is resisted by stresses, or internal forces within the element. The stresses which are
established in components fall into four basic categories: shear stress, tensile stress,
compressive stress and torsional stress.

Shear stress
Shear stress (Figure 1.6) is the internal force created within a structural member
which resists a tendency, induced by an externally applied loading, for one part of
the member to slide past another.

Tensile stress
Tensile stress (Figure 1.7) is the internal force induced within an element which

Figure 1.6
Shear stress.

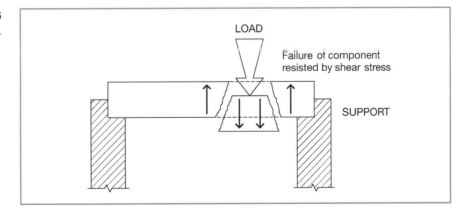

LOAD

Failure of component
resisted by shear stress

SUPPORT

Figure 1.7
Tensile stress.

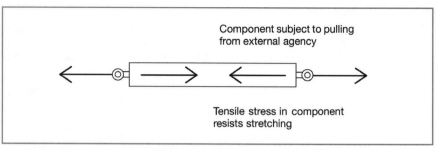

resists an external loading that produces a tendency to stretch the component. When such a force is applied the member is said to be in tension.

Compressive stress

Compressive stress (Figure 1.8) is the internal force set up within a structural component when an externally applied force produces a tendency for the member to be compressed or squashed. Such an element is said to be in compression.

Figure 1.8
Compressive stress.

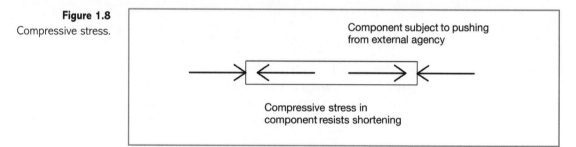

Torsional stress

Torsional stress (Figure 1.9) is the internal force created within a structural element which resists an externally applied loading which would cause the element to twist.

Figure 1.9
Torsional stress.

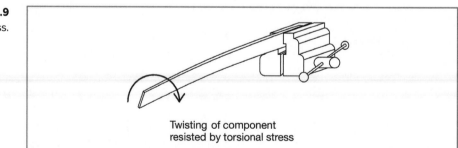

Strain

The effect of a tensile or compressive stress on an element is to induce an increase or decrease in the length of the element. The magnitude of such a change in length depends upon the length of the unit, the loading applied and the stiffness of the material. The relationship between this change in length and the original length of the component gives a measure of *strain* (denoted by *e*):

$$e = \frac{l}{L}$$

where *l* = change in length and *L* = original length.

This effect is also evident in materials subject to shear stress, although the deformation induced in such cases tends to distort the element into a parallelogram shape.

The relationship between stress and strain (subject to loading limits) is directly proportional and is a measure of the material stiffness. The ratio of stress/strain is known as the *modulus of elasticity*.

Moments

The application of a force can, in certain instances, induce a tendency for the element to rotate. The term given to such a tendency is *moment*. The magnitude of such a moment depends on the extent of the force applied and the perpendicular distance between the point of rotation and the point at which the loading is applied. As a result of the effects of leverage, relatively small loads applied at considerable distances from the point of action can induce rotational forces. Moments upon structures must be in equilibrium in order for the structure to remain stable; that is, clockwise moments (+) must be equalled by anticlockwise moments (−). The magnitude of a moment is the product of the force applied and the distance from the point of action at which it is applied (the lever arm) and is expressed in Newton millimetres (Figure 1.10).

Figure 1.10
Moments applied to a building element.

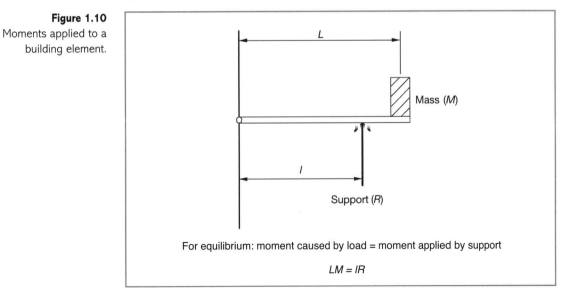

For equilibrium: moment caused by load = moment applied by support

LM = lR

REVIEW TASKS

■ In as few words as possible distinguish between the following forms of stress:

1. Tensile
2. Torsional
3. Shear.

■ Identify examples of where each of these are found in buildings.

■ Consider the shape and form of typical building components such as floor joists, lintels and roof members. How are these influenced by the forces that they are designed to cater for?

■ Visit the companion website at www.palgrave.com/engineering/riley1 to view sample outline answers to the review tasks.

REFLECTIVE SUMMARY

■ *Stress* is the term used to refer to the internal force within an element that is subjected to an external force or loading. Stress may be: shear, compressive, tensile or torsional.
■ *Strain* is a measure of the tendency of an element to suffer deformation under loading.
■ *Moment* is the result of a force applied perpendicular to an element at a distance from a point of rotation.

1.3 | Structural behaviour of elements

Introduction

■ After studying this section you should be aware of the implications of the loads applied to structural building elements.
■ You should understand the terminology associated with the structural behaviour of buildings and the elements within them.
■ You should appreciate the implications of the structural performance upon the selection of materials.

Included in this section are the following areas:

– The nature of forces acting on buildings
– The nature of building components

The nature of forces acting on buildings

Overview

The effects of the types of loads or forces exerted on a building depend on the way in which those forces are applied.

Maintaining the integrity and structural stability of a building relies on its ability to withstand inherent and applied loads without suffering undue movement or deformation. Nevertheless, it is possible to allow for a limited amount of movement or deformation within the building design, as is common in mining areas for example. Resistance to movement and deformation results from an effective initial design of the structure as a unit and the ability of materials used for individual components to perform adequately.

Limited movement, of certain types, is inevitable in all structures and must be accommodated to prevent the occurrence of serious structural defects. The effects of thermal and moisture-induced changes in building materials can be substantial, producing cyclical variations in the size of components. This dictates the inclusion of specific movement accommodation details, particularly when dealing with elements of great size, such as solid floors of large area. Additionally, the period shortly after the erection of a building often results in minor consolidation of the ground upon which it is located; however, this will generally be very minor in nature. These forms of movement and deformation are acceptable; other forms, however, must be avoided. Their nature and extent depend upon the nature and direction of the applied forces. Three main categories of applied forces combine to give rise to all types of building movement.

Vertical forces

Vertically applied forces, such as the dead loading of the building structure and some live loads, act to give rise to a tendency for the structure to move in a downward direction – that is, to sink into the ground. The extent of any such movement depends upon the ability of the building to spread the building loads over a sufficient area to ensure stability on ground of a given loadbearing capacity. The loadbearing capacities of different soil types vary considerably; the function of foundations to buildings is to ensure that the loading of the structure does not exceed the **bearing capacity** or safe bearing pressure of the ground. In most instances the bearing capacity of the ground, normally expressed in kN/m^2, is very much less than the pressures likely to be exerted by the building structure if placed directly onto the ground. The pressure is reduced by utilising foundations to increase the interface area between the building and the ground, thus reducing the pressure applied to the ground (Figure 1.11).

The need to withstand such vertical loads is not exclusive to the lower elements of the building structure, although such loads are greater in magnitude at the lower sections owing to the effects of accumulated loads from the structure. All structural components must be of sufficient size and strength to carry the loads imposed upon them without failure or deformation. Columns and walls, often carrying the loads

Whatever the form and nature of ground in a given location it will have the ability to support loads to some extent. This may be very low or very high. The load in kN that can be supported by 1 m^2 of ground is the **bearing capacity** or safe bearing pressure (SBP).

Figure 1.11
Reduction of pressure
applied to the ground
resulting from the use of
foundations.

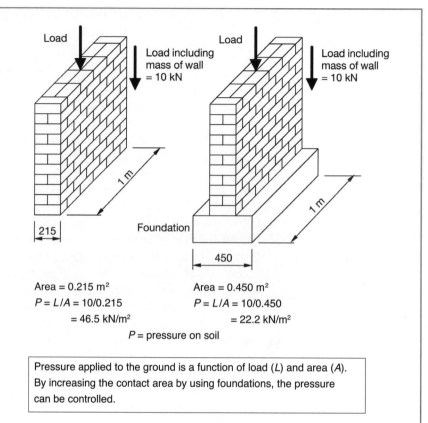

The term **slenderness
ratio** relates to the
proportional dimensions
between height and
thickness of structural
elements.

of floors, roofs and so on from above, must resist the tendency to buckle or to be crushed by the forces exerted. The way in which columns and walls perform under the effects of vertical loads depends on the '**slenderness ratio**' of the component (Figure 1.12). In simple terms, long, slender units will tend to buckle easily, while short, broad units will resist such tendencies. In long, thin components the risk of buckling can be greatly reduced by incorporating bracing to prevent sideways movement; this is termed 'lateral restraint'.

If overloaded significantly even short, broad sections may be subject to failure; in such instances the mode of failure tends to be crushing of the unit, although this is comparatively rare.

Horizontal components, such as floors and beams, must also be capable of performing effectively while withstanding vertically applied loads (Figure 1.13). This is ensured by the use of materials of sufficient strength, designed in an appropriate manner, with sufficient support to maintain stability. Heavy loads on such components may give rise to deflection, resulting from the establishment of moments, or in extreme cases puncturing of the component, resulting from excessive shear at a specific point. When subjected to deflection, beam and floor sections are forced into compression at the upper regions and tension at the lower regions. This may limit the design feasibility of some materials, for example concrete, which performs well

Figure 1.12
Effects of vertical loading on columns and walls.

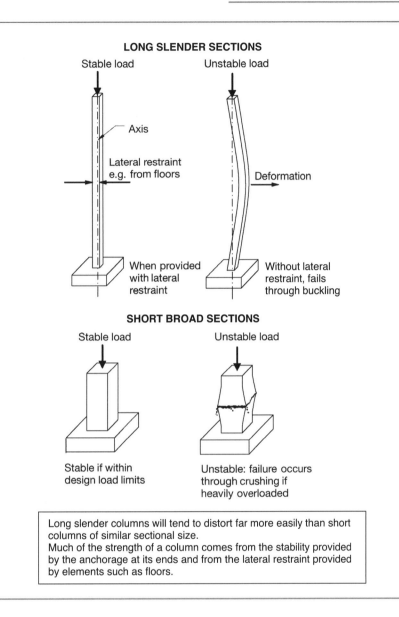

LONG SLENDER SECTIONS

Stable load — Unstable load

Axis

Lateral restraint e.g. from floors

Deformation

When provided with lateral restraint

Without lateral restraint, fails through buckling

SHORT BROAD SECTIONS

Stable load — Unstable load

Stable if within design load limits

Unstable: failure occurs through crushing if heavily overloaded

Long slender columns will tend to distort far more easily than short columns of similar sectional size.
Much of the strength of a column comes from the stability provided by the anchorage at its ends and from the lateral restraint provided by elements such as floors.

Some structural elements are referred to as **composite sections**. They are made up of two or more materials acting in such a way that they maximise the performance of the composite by combining the best characteristics for the individual materials.

in compression, but not in tension. Hence the use of composite units (and **composite sections**) is common, such as concrete reinforced in the tension zones with steel.

As shown in Figure 1.5, illustrating the effects of wind loads, the vertical forces applied to buildings may be in an upward direction. These must also be resisted, usually by making the best use of the mass of the building. Upward loads may also be generated from the ground, in zones of shrinkable clay or those which are prone to the actions of frost, for example. The upward force exerted by the ground in such cases is termed 'heave'.

Figure 1.13
Vertical forces on
horizontal building
elements.

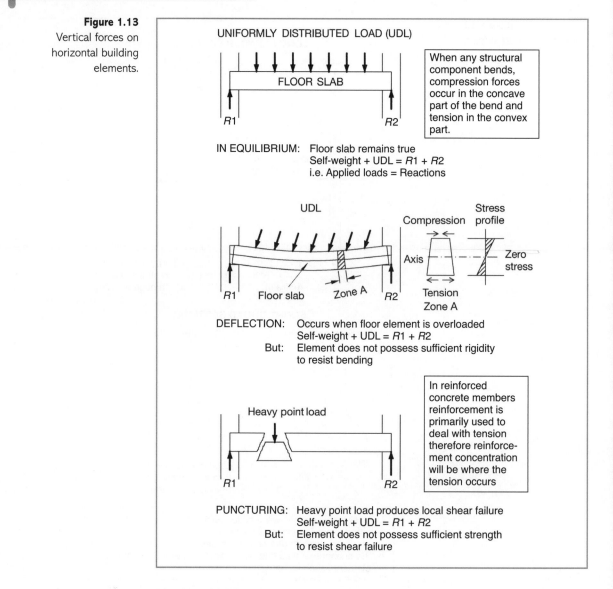

Figure 1.13
Vertical forces on horizontal building elements.

Horizontal forces

Horizontal forces acting on buildings derive from many sources and it is difficult to generalise upon their origins and effects. Typically, however, such loads may be exerted by subsoil pressure, as in the case of basement walls, wind or physical loading on the building. The effects of such forces are normally manifested in one of two ways:

■ Overturning, or rotation of the building or its components
■ Horizontal movement, or sliding of the structure.

These forms of movement are highly undesirable and must be avoided by careful building design. The nature of the foundations and the level of lateral restraint or buttressing incorporated into the building design are fundamental to the prevention

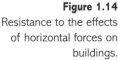

Figure 1.14
Resistance to the effects of horizontal forces on buildings.

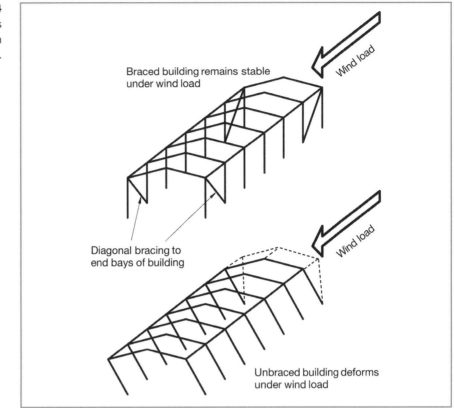

of such modes of failure. Additionally, particularly in framed structures, the use of bracing to prevent progressive deformation or collapse is essential; this could be described as resistance of the 'domino effect' (Figure 1.14).

Oblique forces

In some areas of building structures the application of forces applied at an inclination is common (Figure 1.15). This is generally the case where pitched roofs are supported on walls. The effects of such forces produce a combination of vertically and horizontally applied loads at the point of support. These effects are resisted by the incorporation of buttressing and/or lateral restraint details. The horizontal effects of these loads are sometimes ignored, with disastrous effects.

REVIEW TASKS

■ Explain the value of foundations in dissipating the forces experienced by a wall.

■ Identify examples of horizontal, vertical and oblique forces in buildings with which you are familiar.

■ Visit the companion website at www.palgrave.com/engineering/riley1 to view sample outline answers to the review tasks.

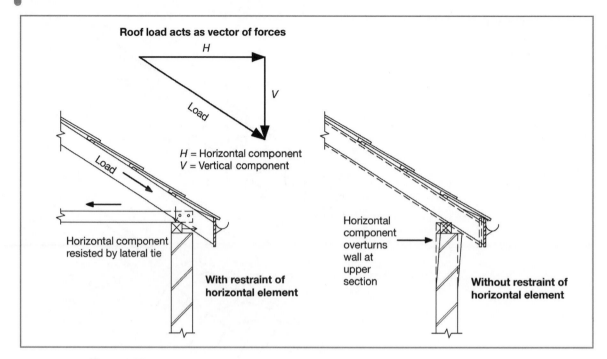

Roof load acts as vector of forces

H

V

Load

H = Horizontal component
V = Vertical component

Load

Horizontal component
resisted by lateral tie

**With restraint of
horizontal element**

Horizontal
component
overturns
wall at
upper
section

**Without restraint of
horizontal element**

Figure 1.15
Effects of oblique loads
on buildings.

The influence of shape – building components

This section is included to extend in a brief way the earlier material concerning the forces experienced by the components of a building. The bulk of the forces experienced by a house would be compression and tension. These tend to translate directly into the use of certain materials suited to these forces – we know that concrete, for example, is excellent in resisting compression and that reinforcement steel is used to counteract tension because of its elasticity. At this stage it may be useful also to consider why the components used in housing are not only of a certain material but also of a certain shape.

We can use the example of a ruler to explain the value of placing material in direct opposition to the forces that they will experience. If the ruler is laid flat and loaded centrally it will bend with relative ease (Figure 1.16).

It will also be seen that if we now rotate the ruler through 90 degrees and load it again, the resistance to bending is dramatically improved, even though its material content is exactly the same. The reason for this is that the body of the material has been placed directly in opposition to the loads to be carried.

This same effect has led to the development of certain shapes of components, such as steel beams. When we think of steel, the I-section generally comes to mind. If we examine the forces experienced by a beam spanning a gap, we know that there is a tendency to bend in a 'smile' fashion. Earlier in this chapter we discussed how this sagging leads to maximum compression at the top of beams and maximum tension at the bottom (Figure 1.13).

Figure 1.16
Placing material opposite
the loads to be carried.

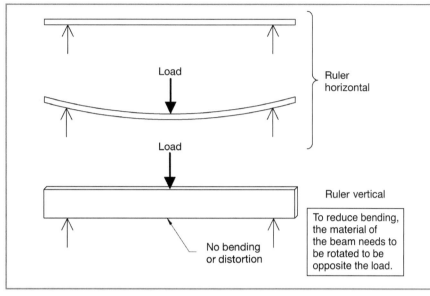

If we now look at the I-shape which has evolved for steel beams, Figure 1.17 shows how the beam has been arranged to place material where it is really needed. The *web* of the beam (like our ruler) is vertical and therefore directly opposes the direction of bending, providing efficiency in resistance to loads. Also, we have just said that when a beam bends, the maximum compression and maximum tension are in the top and bottom of the beam. Here we have the *flanges* of the beam, again placing material just where the forces are to be resisted. This makes the I-section a very efficient shape to use as a beam.

Corrugated materials attempt to place material closer to the vertical plane in order to counteract bending. The effect may be best illustrated by trying to make a flat piece of paper span between your fingers: it sags easily and the paper collapses.

Figure 1.17
The I-section steel beam.

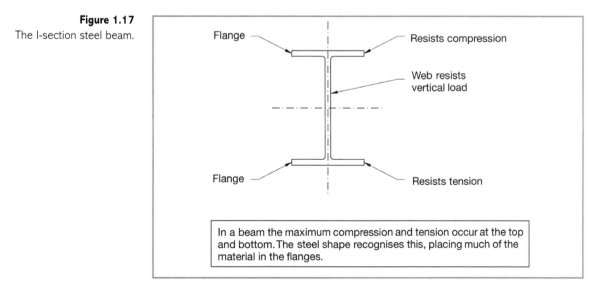

Figure 1.18
Introducing corrugations
to help materials resist
bending.

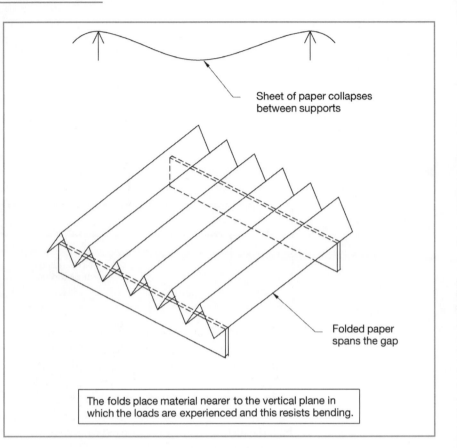

Sheet of paper collapses
between supports

Folded paper
spans the gap

The folds place material nearer to the vertical plane in
which the loads are experienced and this resists bending.

However, if you were to introduce a series of folds as shown in Figure 1.18, the reorganisation of the material and the movement of material towards the vertical help the paper to span the gap successfully.

This principle is found in corrugated roof sheeting and in materials such as wall cladding, which is used extensively on light industrial and industrial premises.

REVIEW TASKS

- We have accounted for the popular sectional shape provided for steel beams. Now consider the shapes of other building elements and consider how their shapes are driven by the loads that they are intended to resist.

- Consider the following key locations in the structure of houses:
  - foundation/wall junction
  - upper floor/wall connection
  - eaves at roof level
  - ridge at roof level.

  What loads/forces are present at these locations and how are the various components and connections designed to transfer or resist these?

- Visit the companion website at www.palgrave.com/engineering/riley1 to view sample outline answers to the review tasks.

2

Preparing to build

AIMS

After studying this chapter you should be able to:

■ Appreciate the criteria involved with the selection of a site
■ Describe the influence of site investigation on project viability
■ Discuss links between the site investigation and foundation design
■ Describe in outline the application of statutory control to the building process
■ Appreciate the extent of utility services and general infrastructure required by a development
■ Outline the preparatory processes that precede formation of the building

This chapter contains the following sections:

INFO POINT

■ Building Regulations Approved Document A, Structure (2004 including 2010 amendments)
■ Building Regulations Approved Document C, Site preparation and resistance to contaminants and moisture (2004 including 2010 amendments)
■ BS12: Specification for Portland cement (1996)
■ BS 4027: Specification for sulphate resisting Portland cement (1996)
■ BS 5930: Code of practice for site investigations (1999)
■ BS 5964/ISO 4463 Building setting out and measurement (1996)
■ BS 7263: Precast concrete flags, kerbs, channels, edgings and quadrants. Precast unreinforced paving flags and complementary fittings (2001)
■ BS 8103: Structural design of low-rise buildings (2011)
■ BS 10175: Investigation of potentially contaminated sites. Code of practice (formerly DD 175) (2011)
■ BRE Digests related to site investigation: 318, 348, 381, 383, 411, 412 and 427
■ *Guidance on Preliminary Site Inspection of Contaminated Land* (1994), Department of the Environment. HMSO

2.1 Selection of sites for building

Introduction

- After studying this section you should be able to discuss the general issues which may be examined before selecting a particular site for development.
- You should appreciate the physical, environmental and financial influences on site selection.
- You should be able to consider the construction of dwellings in the wider context of the development process with regard to statutory restrictions.
- You should be aware of the general implications of the availability and proximity of utility services and associated infrastructure.

Overview

There are many reasons for selecting a site for a new residential development. Typically these include:

- The local demand for housing (which can be actual demand or demand envisaged following consideration of the general growth of the area)
- The proximity of established infrastructure (in the form of roads and services such as sewers)
- The cost of the land
- The quality of land and the effect of this on preparation for use and consequential foundation expense
- The potential for profit
- National, regional and local planning policies.

The decision to select a particular site is generally economic, and the developer will undertake a calculation to assess the likely financial viability of the project. This may be the Developer's equation as outlined below. Remember that if most of the ingredients of the Developer's equation are known, the equation can be rearranged to find out the maximum value of the missing ingredient:

Developer's equation

Income – Outlay = Profit
Income – Profit = Outlay
Income – Profit – Outlay on all items except land = Land cost

The last line illustrates that from the basic Developer's equation we can carry out a residual calculation, looking for the residue that we could afford to pay in this example for the land while still achieving the level of profitability needed.

The local demand for housing will have a marked influence on the profitability element of the Developer's equation. It may suggest the plot area which tends to be demanded by purchasers in the locality. This in turn dictates the number of units that can be built on a particular site, and the local market will define the price that may be charged for them. One of the biggest expenses with any new development is the outlay on infrastructure to be provided – for example, the roads, sewers, footpaths, electricity, gas and water supplies that are likely to benefit the general profitability on groups of properties. Expense on this item will be influenced by the extent of these items already in the locality but also by the costly general nature of essential features like roads through the development. Infrastructure costs are always viewed with some interest, as they generally relate to items of work to be completed or partly executed before building commences on the individual properties. Some time will therefore expire before a chance exists to recover these extremely expensive costs.

The quality of the land in question has always been seen to have a correlation with the nature of the foundations to be provided. The weaker the ground in terms of bearing capacity, the more expensive the foundation solution. Many of the large housebuilders in the UK, such as Redrow, Barratt and Wimpey, have a range of house types that they will build in different locations. Although these house types will be the same from locality to locality, the foundation solution hidden in the ground will vary to suit the conditions of that site. On good ground, variations of simple and relatively cheap strip footings will be provided, while on poor ground the solution may have to be expensive piling. Another issue of growing significance is the development of contaminated land. Here health is the predominant factor and the site will have to be cleansed appropriately; but this of course means an extra development expense, which will in turn affect profit margins. Anything which increases the outlay side of the Developer's equation has an immediate effect on profit.

2.2 | Site investigation

Introduction

- After studying this section you should appreciate the timing of the site investigation in the overall processes of construction.
- You should also appreciate the significance imposed by the findings of the process in terms of material selection, foundation design and overall building form.
- In addition, you should be aware of the range and scope of data that may typically be generated during the desktop study and the on-site investigation process.

Overview

The process of site investigation occurs early in the development of a new project and even before the purchase of the land. This process will reveal a considerable amount of information that often has a major influence on the way in which the building is put together. One of the greatest influences is of course on the foundations to the property: generally, the weaker the ground the more expensive the foundation solution.

When the feasibility of the project is examined the construction detail which arises following the site investigation is usually of great significance. Site investigation is undertaken during the early stages of project development. The Royal Institute of British Architects (RIBA) devised many years ago their 'Plan of Work', which breaks the development process into a series of phases. The earliest of these phases are Inception and Feasibility, which represent the clients having the initial idea of what they want in terms of a building and the cost assessment to ensure that they can afford to pay for the structure.

The plan continues through the phases to Project Completion and Appraisal Feedback. The site investigation happens before the land is purchased during the Inception and Feasibility stages.

The purpose of an investigation may be summarised as follows:

- To assess the ground's composition and characteristics
- To ensure that a foundation is chosen which is compatible with the load to be carried and the strength of the ground
- To ensure safety, efficiency and economy in the design of the building.

Nature of the survey

There are various forms that the site investigation can take, and these range from desktop studies and examination of existing records to reconnaissance by walking over the site, and detailed, extensive testing of samples removed from the site.

The desktop study can supply useful background information on the site, such as may be obtained from maps and historical records about the site. BRE Digest 318 outlines the procedures for undertaking such an exercise and refers to a number of areas for consideration:

- Topography, vegetation and drainage
- Ground conditions
- The proposed structure.

Table 2.1 outlines some of the questions that may be asked of these areas.

It is important to recognise, however, the value of actually visiting the site, and even a simple topographical inspection can be far more valuable than extensive studies conducted in the office. The topography is effectively what you see from ground level: how the land lies, the nature of vegetation and plant growth, the feel of the soil (is it waterlogged, for example?), buildings and natural features.

You may insist on or prefer to select a full-blown investigation, from which extensive quantities of data may emerge. The more that is known about the site, the less any uncertainty about design and performance matters. The size and shape of the site and the proposed size and shape of the building on that site will generally be most influential in the methodology adopted during the undertaking of a site inves-

Table 2.1   Typical site investigation questions.

Topography, vegetation and drainage	Does the ground slope, and if so to what angle? Are there any natural water areas on-site (ponds, streams and so on)? Are there trees or hedges?
Ground condition	What strata exist in the ground? Are these strata typically associated with problems (peat layers, shrinkable clay and so on)? Is ground strength information available? Is there any ground instability or subterranean activity likely to cause instability (such as coal mines or abandoned workings)?
The proposed structure	What area will the building cover? What foundation loading is anticipated? Might the soil create differential movement?

tigation. By the end of the exercise the aim is to compile a representative summary of the features, characteristics and qualities of the site.

Method of extracting information

The ways in which information is obtained about the soil are wide and varied. A traditional technique is to remove samples of ground for site and laboratory analysis by use of a sampling rig and an open drive sample tube.

The samples extracted in this way may be either *disturbed*, where the drilling extraction process breaks down the natural state of the ground because of the auger drill bit used, or *undisturbed*, where the sample taken shows the state of the ground just as it is. Hollow cylindrical tubes with a bottom cutting edge are driven into the ground and take a core sample, as shown in Figure 2.1.

Figure 2.1
An open drive sample tube.

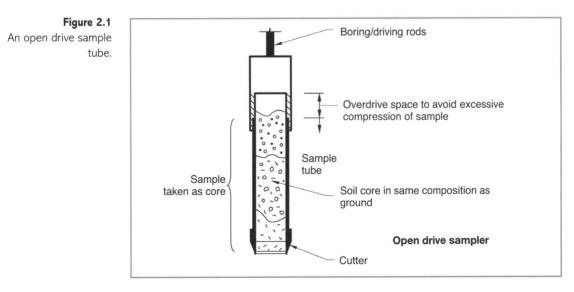

When samples are taken, a borehole log (chart) is used to summarise the findings at different depths in the ground.

In both of the techniques just discussed material samples are removed, but it is possible to obtain information regarding certain features of the ground (such as strength) without removing samples at all (Figure 2.2).

The resistance to rotation under the torque pressures experienced by driving the vanes of the deep vane machine into the ground reflects compressive strength. **Seismic pulses** generated by firing a charge may be used to cause shock waves to bounce back off the various strata in the ground and provide a visual picture of the soil layers present.

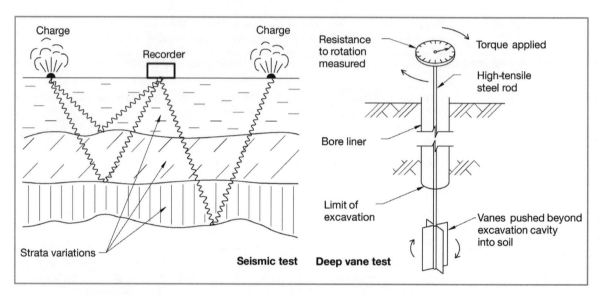

Figure 2.2
Deep vane and seismic methods of ground assessment.

Data generated by a site investigation

A site investigation may provide information concerning:

- The nature of the soil
- The thickness of the layers of different types of soil at the test location
- The strength of the soil
- The existence of contaminants in the soil
- The degree of moisture present
- The existence of a water table
- The location of existing services.

The nature of the soil will concern the basic form that the soil takes. Is it cohesive (for example, clay, capable of moulding between the fingers like plasticine and with very small particle sizes), or non-cohesive (for example, sand, with relatively larger particle size and not capable of moulding under finger pressure when dry)? In reality, the neat division between soil classes in this description is often not found on the site.

Strata composition is the term used to describe the various layers that make up the ground on the site. The ground is generally made up of a number of layers of different soil types, varying with depth.

As well as establishing the classification of the soil, designers will be interested in trying to create a visual picture of the **strata composition** of the ground. When viewing this, we can see how far the loads are likely to penetrate the various soil layers and in particular which layers are likely to be affected.

One of the main features of the ground that site investigations try to establish is its strength, which is usually described in terms of its ability to withstand load or bearing capacity. Strength is measured in kN/m^2, with the range being typically from $50 kN/m^2$ to $600 kN/m^2$, as in Figure 2.3.

Figure 2.3
Indicative bearing capacity.

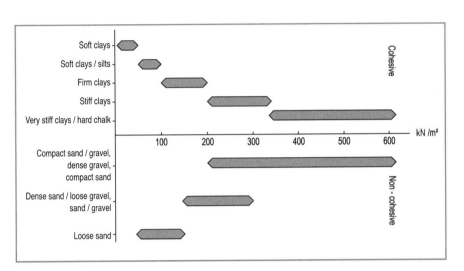

The bearing capacity of the soil is one of the most useful pieces of information as far as the design of the foundation is concerned. Three factors are inseparably inter-related: the load from the building, the bearing strength of the soil, and the area of transmission of the foundation making contact with the soil. Designers need to review the likely load from the building (typically not more than 30 kN/m run of external wall for a two-storey house) and the bearing capacity of the soil, and create a foundation area of transmission which can safely transmit the building load to the ground without exceeding the ground's capacity while providing also for a factor of safety.

The principle of this consideration of three factors can be illustrated as in Figure 2.4.

It would be useful now to refer to Approved Document A of the Building Regulations, Section 2E (foundations of plain concrete), Table 10. This table shows clearly the relationship between building load, ground strength and width of strip footing foundation concrete. You will notice that if the load remains constant but the variable is the strength of the ground, the width of the foundation concrete also varies – the weaker the ground, the wider the concrete needed to spread the load. Similarly, if the ground strength is constant but the load from the building is the variable, the concrete foundation width again changes appropriately.

Another feature of the nature of the soil will be its chemical composition. This may show many impurities, but one of the main contaminants that may be present

Figure 2.4
Load spread through a
simple foundation.

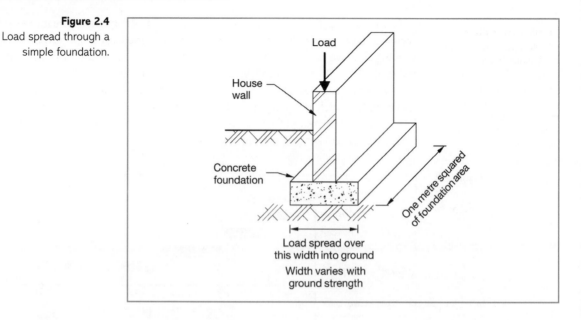

will be *sulphates*. These tend to attack the tricalcium aluminate ingredient of ordinary Portland cement, and when found cause designers to use sulphate-resisting cement to BS 4027 rather than Ordinary Portland to BS 12.

If sulphate attack occurs there is a breakdown of any cement-containing material, with a subsequent loss of strength. Remember that not only will concrete suffer, but also mortar between bricks, traditional mortar joints to drain pipes and so on.

One feature of the ground which may have a big influence over the ease with which construction operations take place on-site will be the existence of water, and more particularly the existence of a water table. The water table represents the upper limit in the ground to where saturation of the soil is occurring, and sometimes this upper limit is close to the ground surface. The existence of water may necessitate temporary groundwater control and the employment of forms of temporary dewatering schemes. It will also influence the general ease with which excavation operations are conducted, together with the cost of those activities.

The construction of a new housing development will generally involve two broad stages: the *pre-contract* stage and the *post-contract* stage. The former represents all the activities that take place before the legal construction contract is signed (such as the JCT Contract) and the latter the activities after the contract is signed. The distinction between the stages is significant from a construction technology perspective in that it marks the change point from design and procurement to the physical construction activity of the project. As the discovery of a water table during the site investigation takes place early in the development of the scheme this is a pre-contract discovery. Later, when construction commences on-site, the water table level will be read again to see whether it has fluctuated, primarily in order that the contractor can be paid an equitable rate for the extent of the excavation in water.

Sometimes there will be existing underground services on the site, such as water mains, sewers and electric power cables. These often have to be re-routed to suit the new development, and awareness of their location is an obvious advantage before excavations into the ground commence.

Contaminated land

Many sites now developed for housing have been previously used; they are referred to as brownfield sites and in some circumstances they are likely to be contaminated. The forms that contaminants take are extensive and include:

- Chemicals – processing chemicals in particular
- Heavy metals, such as mercury and lead
- Fibrous materials linked to respiratory problems, such as asbestos
- Explosives
- Nuclear waste
- Natural radioactivity, such as that caused by radon gas
- Landfill gases, such as methane
- Liquids, such as petroleum.

This list can be easily extended.

The fact that many sites have *contamination* problems is recognised by the Building Regulations in Approved Document C, which makes specific reference to Dangerous and Offensive Substances. This concerns the need to avoid danger to health caused by substances found on or in the ground which is to be covered by the building. Part C of the Regulations also recognises the sensitivity of developments in terms of the use of the site. For example, a car park or hardstanding would not be very sensitive when compared with a housing development. In the event of a certain contaminant of an equal concentration existing on both of these types of site, the degree of corrective work might be quite different. The name given to remedying contaminated land is *remediation*.

In order to fully assess the potential for contamination to pose a risk to the development and to people or eco-systems, a structured approach to 'risk assessment' is necessary. Typically this will follow the structure set out below.

- *Hazard identification*: preliminary site assessment based on desktop study and walkover
- *Hazard assessment*: more detailed investigation to assess potential pollutants and linkages
- *Risk estimation*: estimating scale of risk and possible consequences
- *Risk evaluation*: assessing whether risk is acceptable or not acceptable.

The general approach to risk assessment will be based on a conceptual model for the site which defines sources of contamination, possible pathways and potential receptors (Figure 2.5).

As investigating contamination is potentially more important to health than ordinary site investigations, the approach taken in the evaluation of the site might be quite different. There is a British Standards Institution reference which may be reviewed in this area, namely BS 10175. The fact that this standard was in existence

Figure 2.5
Conceptual
contamination risk
model.

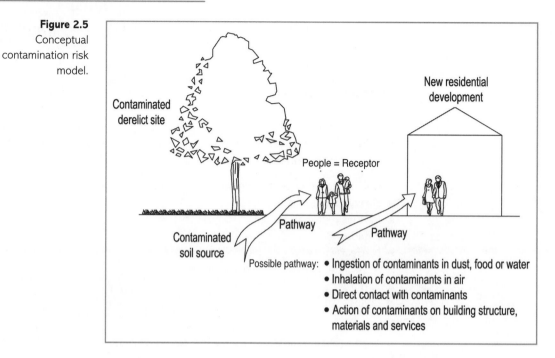

as a Draft for Development (DD 175) from the mid-1980s and only achieved full BS status in 2001 is probably a reflection of the difficulty of compiling a definitive approach to researching contamination problems.

REVIEW TASKS

■ Name *five* typical results that you would expect to emerge from a site investigation and rank these into importance, justifying your ranking.

■ Identify *two* potential sites for building an estate of 30 houses near your home: one a new or greenfield site and one on a previously used brownfield site. Generate a list of the key differences between the two in terms of construction difficulties. How would these be evaluated using site investigation?

■ Consider the content of a typical Environmental Impact Assessment for the above sites. What would the main issues be?

■ Visit the companion website at www.palgrave.com/engineering/riley1 to view sample outline answers to the review tasks.

2.3 | Overview of statutory control of building

Introduction

- After studying this section you should be able to appreciate the evolutionary development of building control through public health and other associated legislation.
- You should be able to outline the basic process of obtaining formal approval to build, and appreciate how control extends to the building process.

Overview

The application of controls over the building process is a relatively new concept in the UK. Prior to the first national set of Building Regulations in 1965, control over the building process was limited. The first real major realisation of need for control probably came in the aftermath of the Great Fire of London in 1666. The rate of spread of the fire itself was a major factor in the disaster and this was a consequence of the high density of building, with few open spaces, coupled with the combustible nature of the building materials of the time.

Following the fire, control regulations were issued which were applicable to London itself.

A main feature of the development of control in the rest of England was the various Public Health Acts (PHA), and probably the most notable of these was the PHA 1875. This had three main focuses: structural stability, dampness and sanitation. Following this Act, a set of Model Bye-Laws was issued in 1877 as a guide for local authorities, who were delegated the responsibility for setting and enforcing minimum standards of construction. The concerns of the bye-laws were primarily health and safety, and this set of guidance controls largely followed the contents of the Public Health Act of two years earlier.

Following the limited progress in the 19th century, legislation concerning construction was also limited and mainly in the area of public health, for example the Public Health Act 1936. By the 1950s many local authorities were issuing bye-laws peculiar to their locality, and this made life difficult for the construction professional, in that laws varied from area to area. Because of the need for consistency, the National Bye-Laws were established in 1952. Following these, items of significance include the Public Health Act 1961 and the first national set of Building Regulations in 1965.

Building control

The aim of the Building Regulations was to set what are considered to be the minimum standards that are acceptable in construction. Despite the significant step forward in national construction standards as a result of these Regulations, some considerable difficulty was being experienced in ensuring that they were enforced.

Consequently, in 1972 the Government established the Local Government Act, which clearly set the responsibility of Building Regulation enforcement with the local authorities, and this in turn forced these authorities to employ personnel specifically for this purpose.

Since then, enforcement has been carefully controlled, and this is still the situation today. Following the first set of Building Regulations, refinements were forthcoming which were issued initially as amendments and eventually incorporated into the body of the Regulations with the issue of the Regulations in 1972, 1976, 1985 and 1991.

In 1985 the format layout of the Regulations changed substantially to the A4 format still used today. As in the past, the Regulations tend to divide the building into a number of logical sections: Structure, Fire, Ventilation, Stairs and so on. Since 1985, each of these sections has been enhanced by the use of an Approved Document. These explain the application of the Regulations in detail and contain the appropriate British Standards that apply. Section 3 of the Building Regulations lays down the procedure to be followed to obtain formal approval to build, and contains sections such as Notices and Plans, Control of Building Work, and Relaxation of Requirements.

The Building Regulations (for England and Wales) are split into a series of approved documents that broadly reflect the performance requirements of buildings as described earlier. They are as follows:

- **Part A**: Structure – this sets out requirements to ensure that the building will cater for anticipated loadings safely.
- **Part B**: Fire safety – this has the aim of controlling both the materials used according to the degree of risk and making sure that buildings can be evacuated without loss of life (means of escape) in the event of a fire.
- **Part C**: Site preparation and resistance to contaminants and moisture – control of design and construction to cater for water penetrations and condensation in buildings.
- **Part D**: Toxic substances.
- **Part E**: Resistance to the passage of sound – deals with passage of sound through walls and floors between dwellings.
- **Part F**: Ventilation – sets out requirements for the ventilation of habitable rooms and unheated voids.
- **Part G**: Sanitation, hot water safety and water efficiency – lays down requirements for sanitary appliances and drainage.
- **Part H**: Drainage and waste disposal – this is aimed, primarily, at meeting health and environmental standards.
- **Part J**: Combustion appliances and fuel storage systems – a key element of this section deals with the safe discharge of flue gases.
- **Part K**: Protection from falling, collision and impact – this deals with requirements for the design of stairs, ramps and vehicle barriers.
- **Part L**: Conservation of fuel and power – this aims to ensure that the building is properly insulated and built to minimise environmental changes.
- **Part M**: Access to and use of buildings – this aims to ensure that appropriate access is available to buildings and facilities for disabled people.

- **Part N**: Glazing – this aims to control the use of and safe positioning of glazing in windows, doors and so on.
- **Part P**: Electrical safety – this deals with the safe installation of electrical systems.

It is not appropriate in an introductory text such as this to explore each of these areas in detail, however the principles that are set out in the Regulations are reflected in the individual sections of the book as appropriate.

Town and country planning

When a person wishes to obtain formal approval to build, there are broadly two areas of concern: compliance with the Building Regulations in terms of design and material content, and the suitability of the building for the location and the piece of land in question. The latter area draws in the Town and Country Planning Acts, which govern the use of land throughout the UK. Land is classified for different uses (for example, residential use or agricultural use), and these pieces of legislation restrict the building to forms designated as suitable for the particular site.

Typically, planning permission may be required if you want to:

- Extend a flat externally
- Create self-contained accommodation from part of an existing house
- Divide off part of a house for business purposes
- Erect something that may restrict the view of road users (such as a fence)
- Build in a way which was not included in the original Planning Permission for the house
- Widen access or supply access to a road.

Minor changes to houses may be undertaken without the need to apply for planning permission under *Permitted Development Rights*; for example, where the proposed extension has a volume which is less than 15 per cent of the original house volume (calculated on external measurements) or 70 m^3, whichever is the greater. The situation can, however, be complicated if certain conditions apply, such as the property being located in a conservation area, or where other extensions have already been added.

Planning permission may also be required for putting up certain fences, walls or gates, or forming patios, paths and driveways.

The planning system in England and Wales has been subject to recent reforms, largely associated with the concept of 'localism' with the adoption of a 'Local Plan' as the key development planning document prepared by local planning authorities. A key feature of the Local Plan relates to the local need for jobs and homes. As such, there is a significant linkage to the development and construction of dwellings. It is intended that local planning involves and is informed by active engagement with the community through the use of 'Neighbourhood Plans'. In addition, there is strong reference to mitigation of, and adaption to climate change. For example, there has been significant development in the approach to construction in areas that may be at risk of flooding.

Other forms of approval may also be needed in connection with contemplated works:

- Listed building consent
- Conservation area consent
- Where tree preservation orders exist
- Any work that is undertaken that may affect nearby or adjacent buildings, as structure may be affected by the Party Wall etc. Act 1996. The general principle of this Act is to enable building owners to undertake specified works on or adjacent to adjoining properties while affording protection to those that might be affected. This is particularly relevant to refurbishment and alteration works.

Health and safety

A major consideration on any housing project is the safety, health and welfare of all those associated with the building process (such as construction workers, subcontractors) and those affected by the building operations (such as visitors to the site, passers-by). Such issues are controlled by legislation which, largely, sets the standards to be achieved rather than prescribing detailed rules to follow. The **Health and Safety at Work etc Act 1974** is the primary legislation that imposes a **duty of care** on employers (such as developers, contractors) to safeguard employees and others affected by their operations (such as a housing development). The Act also imposes a duty of care on workers for their own safety and that of fellow workers. Employers must provide a safe place of work and safe methods of working and those with more than four employees are required to have a **health and safety policy**. The Act enables other legislation to be enacted and there is a large body of subordinate legislation that affects development and construction (such as CDM, Work at Height, Lifting Operations, Work Equipment, COSHH and so on). The focus of UK health and safety legislation is the **management of risk** and, as such, hazards (such as contaminated ground, deep excavations, roof work) have to be identified, eliminated or controlled both during the design stages and when the construction work gets underway. The management of risk is a statutory requirement under the **Management of Health and Safety at Work Regulations 1999**. This subordinate legislation is not construction specific and it applies to all employers irrespective of industry. In a construction context 'the Management Regulations' apply to developers, architectural practices, contractors, subcontractors and so on, irrespective of size. Employers with five or more employees must have a formal procedure where risk assessments are written down, reviewed and revised appropriately. This applies equally to designers and contractors. Further provisions of 'the Management Regulations' require appropriate arrangements to be made *'for the effective planning, organization, control, monitoring and review of the preventative and protective measures'*. In practice this means the provision of a safety management system including safety inductions for workers and visitors to site, the preparation of method statements for high risk work, consultation with the workforce and audits of health and safety management performance. The **Construction (Design and Management) Regu-**

lations 2007 are construction-specific legislation that applies to 'construction work' including site investigations, demolition and site clearance, infrastructure such as roads and drainage, new build construction, alteration, renovation, repair and maintenance and so on. CDM establishes:

- General duties to manage health, safety and welfare
- Further duties for notifiable projects (the HSE must be notified where the construction phase is likely to last for more than 30 days or involve more than 500 person days of work)
- Specific duties for health, safety and welfare provision on-site.

CDM duties are imposed on a variety of 'duty holders':

- **Clients** who commission or carry out construction work
- **Designers** such as architects, architectural practices, contractor/developers, design and build contractors, clients who prepare in-house designs
- **Contractors** who are in the business of carrying out or managing construction work such as contractor/developers, main contractors and subcontractors.

Clients must ensure that all designers, contractors and others they appoint are competent to undertake their statutory duties. They must also provide such information as is available to enable them to carry out these duties; this forms part of the **pre-construction information**. Furthermore, clients must ensure, as far as they are able, that the health, safety and welfare arrangements on-site comply with the Regulations. CDM duties apply to all designers (architects, structural engineers and so on) irrespective of the size of project or whether the project is notifiable or not. They must notify clients of their statutory duties, make sure that they 'risk assess' their designs to avoid hazardous working on-site and convey information to clients, other designers and contractors so as to enable them to comply with their own duties under CDM. For notifiable projects, an important part of the client's duties is the appointment of two further 'duty holders':

- A **CDM coordinator** who gives advice and assistance to the client, designers and contractors, collects and disseminates the pre-construction information to all designers and contractors, and makes sure that health and safety issues are addressed during the design stages of a project
- A **principal contractor** who, amongst others things, is responsible for preparing a **construction phase plan** for managing the health, safety and welfare regime on-site.

Contractors also have CDM duties and these include managing their activities on-site so as to avoid risks to health and safety. This means carrying out safety inductions, communicating the site rules to workers, communicating information to workers via tool-box talks and preparing and communicating method statements to ensure safe working. On notifiable projects, all contractors must cooperate with the principal contractor by following directions, complying with site rules and providing information such as risk assessments and method statements.

At the conclusion of the construction phase of a notifiable project, the CDM coordinator must hand over a **health and safety file** to the client. This contains all sorts of information regarding the completed project so as to alert people to possible risk when undertaking future maintenance, cleaning work, alteration or demolition, and so on. Preparation of the health and safety file involves liaising with the principal contractor regarding the format of the file and collecting information from the client, designers and contractors. Health and safety legislation is part of the UK criminal code and breaches can lead to heavy fines and, in serious cases, custodial sentences for offenders. Additionally, companies and organisations can be prosecuted for health and safety management failures under the **Corporate Manslaughter and Corporate Homicide Act 2007** where there is a gross breach of a duty of care.

Environmental legislation

The Environmental Protection Act 1990 (EPA) and the Environment Act 1995 have had a major impact on procedures on-site and also in respect of legal responsibilities related to site pollution. The EPA has also dealt with the issue of hazardous waste disposal and has established the present Landfill Tax system which provides for dangerous waste material to be handled by licensed tips. As there is a charge per tonne for disposal of the hazardous waste this may produce a considerable expense, which could even affect the viability of projects. Remember that, in terms of the Developer's equation, if there is an increase in development costs this has a direct influence on profit.

REVIEW TASKS

- What *two* processes are normally undertaken when seeking formal approval to build from a local authority?

- What are the *CDM Regulations* and when do these apply?

- Generate a chronology of environmental legislation affecting house construction in your country.

- Visit the companion website at www.palgrave.com/engineering/riley1 to view sample outline answers to the review tasks.

2.4 | Overview of utilities and infrastructure

Introduction

- After studying this section you should be able to explain the content and significance of infrastructure in the development of dwellings.
- You should be aware of the entry details of utility services to individual properties.

Overview

Utility services supplied to a residential development relate to the services of water, electricity, gas and sewers. The term 'infrastructure' also includes items such as roads, footpaths and street lighting. These items represent early expenditure on the project, as they are installed or completed before work really commences on individual house units. The extensive cost associated with them is generally divided among the house units when setting retail prices. The provision of utility in the UK was once the responsibility of public sector utilities providers. We now have a privatised system that allows competition between alternative providers. However, they all utilise the same national utilities infrastructure.

Potable water is water that is treated to ensure that it is free from contaminants and biological agents that may be harmful to health. In simple terms it is 'drinkable'.

Cold water supply

It is a requirement of the Water Act that every dwelling should be provided with a **potable water** supply. On its way to a house the water passes through various processes and often along extensive pipelines. Figure 2.6 shows this route and indicates the pressures needed to overcome resistances in the pipe and fluctuations in

Figure 2.6
Water supply national distribution.

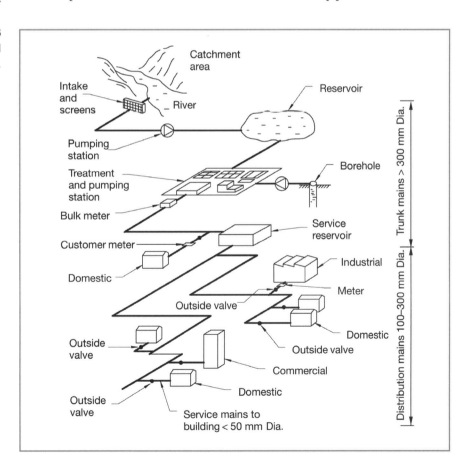

demand, ensuring that sufficient pressure is available for correct delivery. The water provided may come from a variety of sources, including reservoirs and underground wells.

The water main supplying residential developments is normally located in the public footpath outside the boundary of the house units. At this location a water company stopcock is normally positioned to allow the water supply to the individual house to be cut off without the need to enter the boundary of the property. From this point the water pipe supplying the house is termed the *cold water service pipe* (Figure 2.7). As shown in the figure, this service pipe is buried some distance in the ground to avoid damage and prevent freezing in cold weather. In the UK the ground rarely freezes beyond 600 mm depth so at 1 m depth this pipe should be safe.

Figure 2.7
Cold water service pipe.

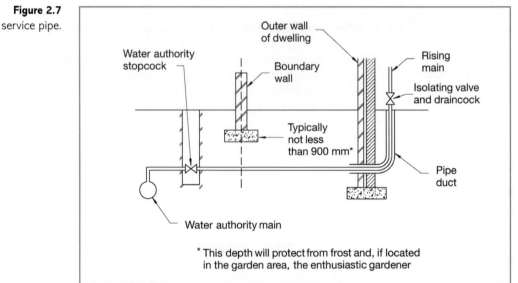

The service pipe was traditionally made of lead but it was subsequently realised that lead may dissolve into the water and affect health, hence the pipe is now made of high-density polythene. To assist in identification on site, this pipe is bright mid-blue in colour. As the pipe is flexible it will assist entry to the property if a duct of drain pipe material or similar is used through the external wall and extended to the top of the ground floor finish to allow the pipe to be threaded through.

Once this pipe emerges beyond floor level a stopcock is used to allow the occupier to close down the cold water supply in the event of a pipe burst or while undergoing maintenance work. On leaving the stopcock this cold water supply pipe is referred to as the *rising main* as it moves up through the property, historically terminating at the cold water storage tank. However, in more modern installations cold water storage is unlikely to be present; instead all cold water taps and the heating boiler are fed at **mains pressure**.

Mains pressure is the pressure that the water is transmitted at in the main feed to the property. In domestic situations this is normally not less that 1 bar – this is the pressure needed to lift water to a height of 10 m.

Electricity supply

As with the supply of water, an established delivery network is the means for conveying electricity supplies to residential developments. Again, as with the water supply, resistance has to be overcome and sufficient capacity provided to ensure adequate supplies to the consumer. This time the resistance is, of course, electrical resistance rather than friction.

To ensure that the correct strength of supply is provided to the householder, the National Grid for distribution has to carry some extremely high voltages, as shown in Figure 2.8. As we get nearer to the point of use a series of reductions in voltage is ensured by the use of transformers. In the footpath to a residential development we will generally have a three-phase four-wire supply: three live wires and one neutral. Between each of the live wires the voltage potential is now 415 V and between any of the live wires and the neutral wire 240 V. This means that to tap into the cable in the footpath outside the boundary to the house we need to connect to one live (phase) and the neutral to connect the property (Figure 2.8). This 240 V supply is termed single phase, as only one live wire is used.

The underground wires are wrapped in protective covers as they are brought to the house. Modern housing has a meter box on the outside wall, accessible from outside the premises for meter-reading purposes and similar to those illustrated later for incoming gas supplies.

Figure 2.8
House and tapping into a three-phase supply.

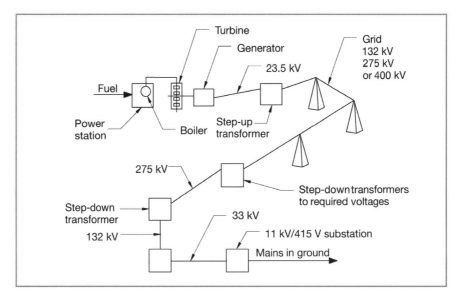

Gas supply

Gas was historically generated by processing coal and was referred to as 'town gas'. This was not a very environmentally friendly fuel because of the carbon placed in the atmosphere during production. However, most of the gas consumed in the UK is now natural gas, methane from natural tapped pockets buried below the sea or ground. This natural gas has a much higher calorific value (heat produced per unit), but requires more air to assist in the combustion process than its old rival. In terms of national distribution (Figure 2.9) there is now an established pipework network and, similar to moving water through pipes, the gas requires significant pressures to ensure that a constant supply is available.

Figure 2.9
National distribution of gas.

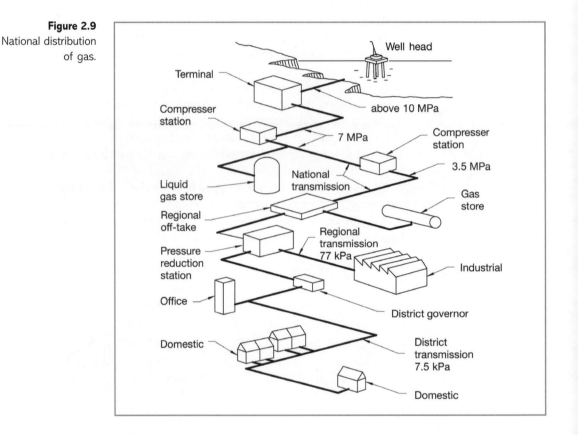

High-pressure supplies of course need to be reduced as they move towards the final consumer, and even the house gas meter has its own pressure governor fixed to the inward connection of the gas meter.

Supplies of gas tend to be conveyed in high-density plastic pipes, as with the water supply, but this time are coloured yellow for identification purposes. The supply main is often in the pathway or road beyond the house boundary, and the supply is brought in below ground. On modern house developments a meter box is typically used of the same type as for the incoming electricity supply. Figure 2.10 shows a common way to bring the supply into the premises, and ducts are often used as penetration of the wall is achieved to protect the supply from damage.

Having services of different types in close proximity as they are routed to a development may be an advantage, but it can create some difficulties; damage when gaining access is not uncommon.

Figure 2.10
External gas meter box.

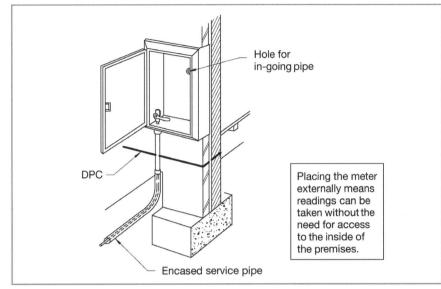

Hole for
in-going pipe

DPC

Placing the meter
externally means
readings can be
taken without the
need for access
to the inside of
the premises.

Encased service pipe

Drain connections

Waste water is the term
used for any water that
has passed through the
building for any use such
as washing, sanitary
appliances, and so on.

Surface water is
rainwater and is collected
into drains and gullies to
remove it from the
immediate environs of the
building.

Below-ground drainage is provided to move **waste water** and **surface water** from the property and its immediate vicinity, and pick-up points for this are generally located around the outer perimeter of the house. Many of these drain runs begin with a gully – for example where rainwater discharges from downpipes to the drain or where waste pipes leave the property containing discharges from sanitary appliances, such as kitchen sinks, baths and washbasins. Figure 2.11 shows a drainage layout typically used. Occasionally discharges need to be provided from an internal ground floor toilet (WC) which is inside the external walls of the property, and such connections require the drain pipes to be brought in and set into the ground floor long before the toilet itself is positioned.

Drains are generally connected to a sewer which will carry the discharges to a treatment plant.

Water discharges from housing tend to be divided into two types: surface water discharges and foul discharges. Surface water is rainwater collected from the roof of the property, which by its nature needs no treatment as such. By contrast, foul discharges are those which originate at a sanitary appliance (bath, basin, sink, shower or toilet), and clearly these need treatment at a treatment plant. Recognising the benefit of separating the rainwater from the foul water in terms of the total volume of material requiring treatment, it is best that a *separate system* of drainage is employed. Here all the rainwater collected goes into one set of drain pipes and all the foul water into another set of drain pipes. Each of these pipe systems then discharges into a separate sewer, one for rainwater and another for the foul water.

Some new developments connect to existing sewers. Where this is the case and where only one sewer exists, both rainwater and foul water will be discharged into this one sewer. Here drains surrounding the building mix rainwater and foul water together and are referred to as a *combined system*, as shown in Figure 2.11.

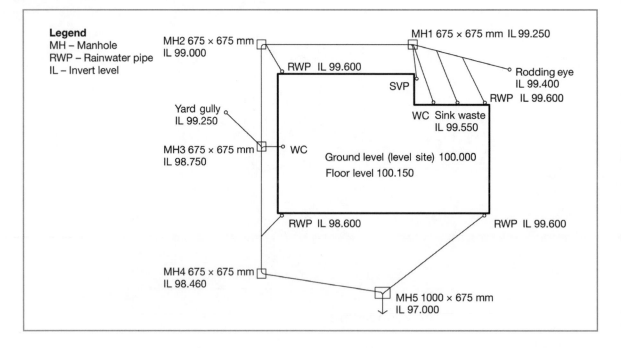

Figure 2.11
Combined drainage
system.

In Figure 2.11 the designation IL means 'Invert Level'. This is the lowest point on the interior bore of the pipe measured at any location in the drainage system. Smaller values of invert level indicate that the pipe is getting deeper into the ground as the dimensions draw closer to the base level of 0.00 m, which is the reference level from which all heights are measured.

Shared access, roads and footpaths

Infrastructure items include shared facilities, such as those provided for access. Roads to residential developments are generally partly formed for access during the construction operations and only finished at the very end of the project. The sewers are located below the various layers of material used for the body of the road. Where the development has two sewers – surface water and foul water – the rainwater collected by the road drainage will generally discharge into the surface water sewer. The only compromise to this is the *partially separate system*, where rainwater from roads is collected by the foul sewer and only rainwater from house roofs goes into the surface water sewer.

Excavation for the road will suit the prevailing ground levels and slopes. Many access roads are cambered and slope from the centre of the carriageway to the kerbs on each side. Others slope in one direction only for directing rainwater off the surface, and this of course means road gully locations to one side of the carriageway only. Kerbs set in concrete to the edges of the road locate its boundary during the laying of the materials for the road body. These have shapes which comply with BS 7263.

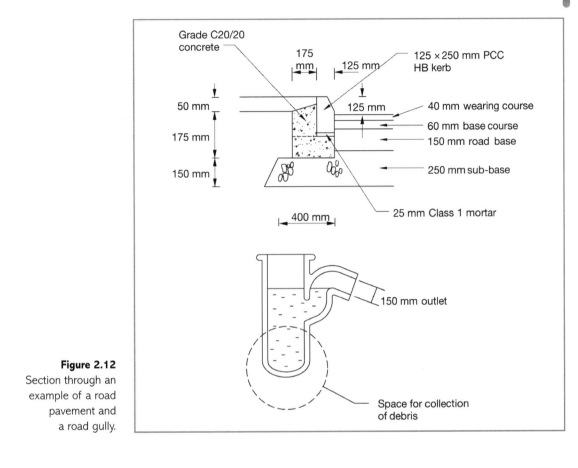

Grade C20/20 concrete

175 mm

125 mm

125 × 250 mm PCC HB kerb

50 mm

175 mm

150 mm

125 mm

40 mm wearing course

60 mm base course

150 mm road base

250 mm sub-base

25 mm Class 1 mortar

400 mm

150 mm outlet

Space for collection of debris

Figure 2.12
Section through an example of a road pavement and a road gully.

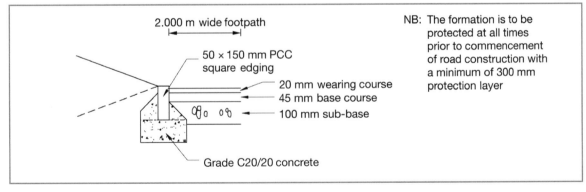

2.000 m wide footpath

50 × 150 mm PCC square edging

20 mm wearing course
45 mm base course
100 mm sub-base

Grade C20/20 concrete

NB: The formation is to be protected at all times prior to commencement of road construction with a minimum of 300 mm protection layer

Figure 2.13
Section through a tarmacadam footpath.

Tarmacadam is the standard surface for road pavements. Figure 2.12 illustrates that this material tends to be applied in layers with a base course underlying a wearing course. Footpaths may also be laid in tarmac on a hardcore (stone) base (Figure 2.13).

Various alternative forms of footpath are illustrated in Figure 2.14.

Figure 2.14
External paving details.

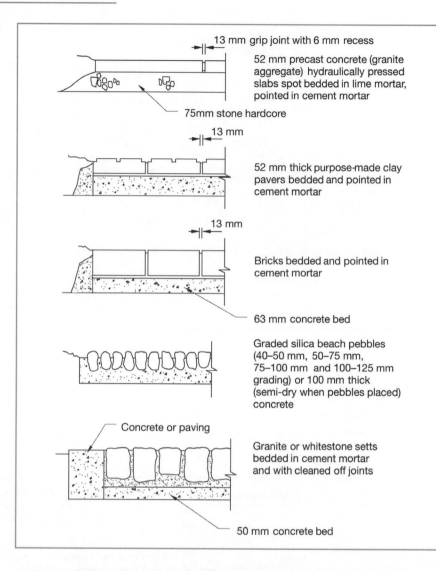

13 mm grip joint with 6 mm recess

52 mm precast concrete (granite aggregate) hydraulically pressed slabs spot bedded in lime mortar, pointed in cement mortar

75mm stone hardcore

13 mm

52 mm thick purpose-made clay pavers bedded and pointed in cement mortar

13 mm

Bricks bedded and pointed in cement mortar

63 mm concrete bed

Graded silica beach pebbles (40–50 mm, 50–75 mm, 75–100 mm and 100–125 mm grading) or 100 mm thick (semi-dry when pebbles placed) concrete

Concrete or paving

Granite or whitestone setts bedded in cement mortar and with cleaned off joints

50 mm concrete bed

REVIEW TASKS

- What colours are used to identify gas pipes and cold water pipes?

- How does the type of material carried by below-ground drain pipes lead to a classification system for the pipes?

- What is the purpose underlying the move away from combined drainage systems towards separate systems?

- Examine the visible elements of the drainage system to your own home (gullies, rodding eyes, inspection chambers and so on) and try to decide whether it is a combined or separate system.

- Visit the companion website at www.palgrave.com/engineering/riley1 to view sample outline answers to the review tasks.

2.5 | Preparing the site for use

Introduction

- After studying this section you should be able to appreciate the typical stages that occur when excavation is undertaken for a property.
- You should understand the content and significance of Part C of the Building Regulations.
- You should be able to identify some of the works which may be used to improve the quality of the site, either permanently or temporarily.

PART 1

Overview

On sites where contamination is not an issue there will be other preparatory works to do before the building can commence, and this is also recognised by Building Regulations Part C. Clause C1 refers to the need to have the ground covered by the building reasonably free of vegetable matter. This will mean stripping the site of growing vegetation, but also it will necessitate the removal of topsoil.

Most sites in the UK will naturally have a layer of topsoil of varying thickness overlying the ground strata below. This topsoil layer is aerated and loamy and ideal for the growth of plant life and grass. We would not want plants to be growing below the building, but another important reason for the removal of topsoil is the fact that it has little if any bearing strength. If we placed the structure directly onto this layer

Figure 2.15
Identifying stages in excavation.

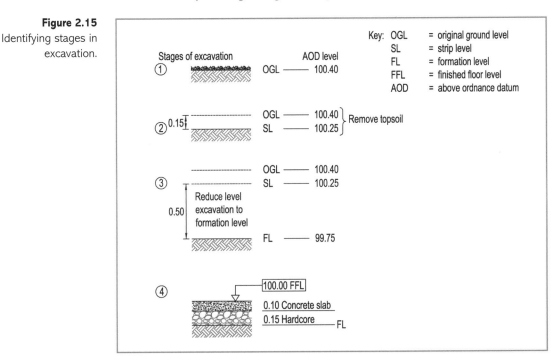

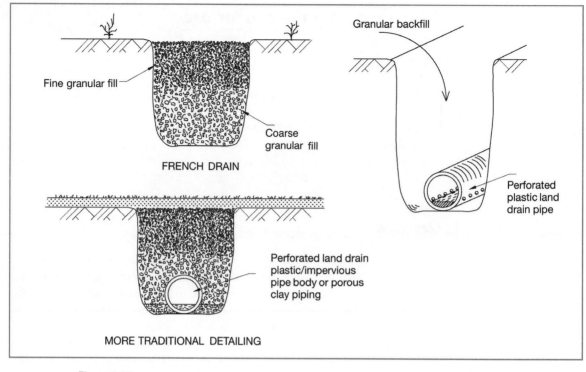

Figure 2.16
Land drainage trenches.

it would sink. The thickness of the topsoil layer will be one of the features provided by the site investigation process, and typically we would expect 150–300 mm. Once removed, the normal procedure would be to store the soil on-site in spoil heaps for later reuse on planting and grassed areas.

Figure 2.17
The moat system of land drainage.

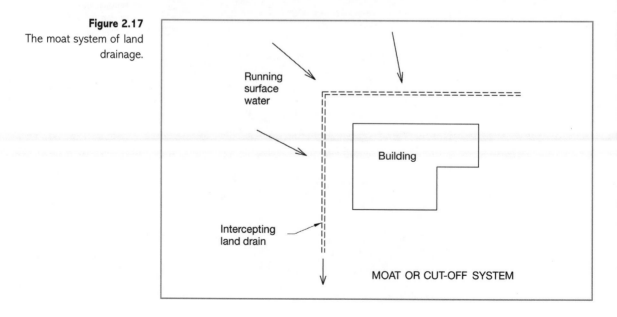

Figure 2.18
Some land drainage
system layout options.

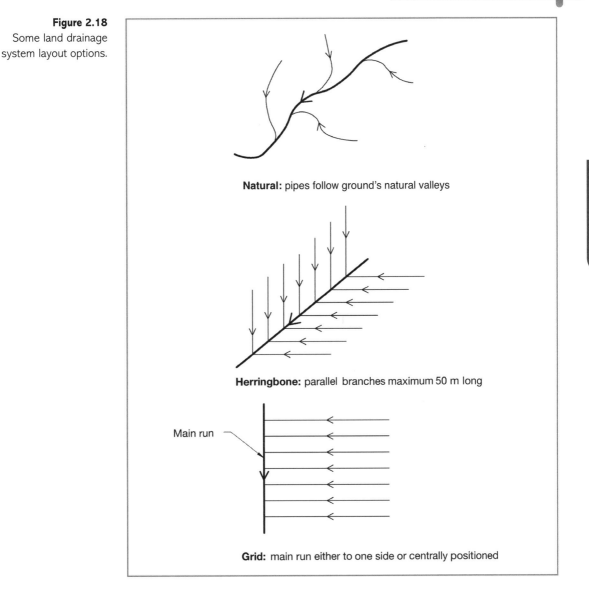

Natural: pipes follow ground's natural valleys

Herringbone: parallel branches maximum 50 m long

Main run

Grid: main run either to one side or centrally positioned

When we start to excavate the topsoil we will be usually commencing at Original Ground Level (OGL), and when we have finished we are at Strip Level (SL). In the earthworks undertaken for the construction of a typical house there are a number of terms and abbreviations used to identify the various levels reached, and these ordinarily come into consideration when dealing with the excavation needed for floors. However, we will illustrate these at this stage as a preamble to the section dealing with floors.

Designers will generally specify the finished level that is needed for the internal floor of a building, and this will be by reference to the national system of levels that is used in the UK. This system uses its base level as 0.00 m, which is mean sea level at Newlyn in Cornwall. The system is referred to as Ordnance Datum. All levels of

Figure 2.19
Soakaway and catchpit details.

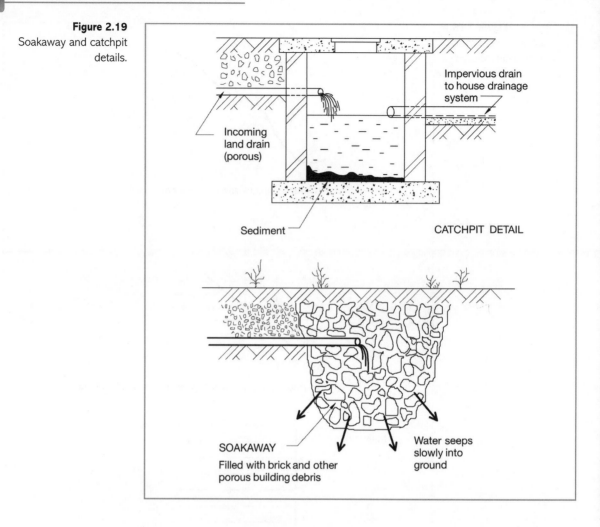

ground and buildings throughout the UK are relative to this level and give the height in metres (previously feet) above this level. In Figure 2.15 the internal Finished Floor Level (FFL) is set as a result of consideration of the general levels of the site being developed and mindful of the fact that we require the internal floor to be above outer ground level to help protect the building from moisture entry. This is also the subject of the Building Regulations, Part C, which requires the floors, walls and roof to resist moisture entry.

Part C of the Building Regulations refers to the possible need to apply subsoil drainage to lower ground moisture levels, protecting the fabric of the building in the ground and reducing the chance of the movement of moisture into the building. Subsoil or *land drains* are quite different from the drain pipes that we use to carry the discharges of the house's sanitary appliances, particularly in the fact that they are either porous or perforated. Clearly, with the discharges from sanitary appliances we have to use impervious pipes to contain the material as it is carried to the point of discharge.

Figure 2.20
Well point dewatering
system.

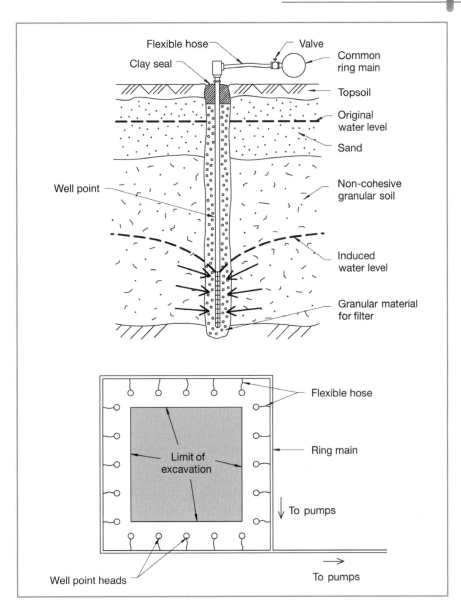

PART 1

Land drainage trenches (Figure 2.16) are generally filled with granular material rather than earth to speed the passage of water down to the collecting pipework. An illustration of the application of this type of drainage is shown in Figure 2.17. Here the house is to be placed at the foot of some sloping ground and the aim of the drainage system is to intercept water running off the face of the sloping ground during or following rainfall.

The pipe layout for the land drainage system may vary from one which follows the natural valleys and contours of the land (natural) to others devised to collect efficiently from the shape of the site in question. Figure 2.18 shows some of the options.

Figure 2.21
Steel sheet piled
cofferdam.

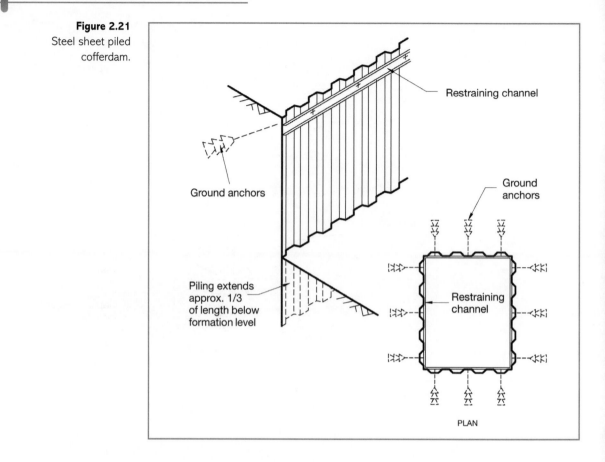

Restraining channel

Ground anchors

Ground anchors

Piling extends approx. 1/3 of length below formation level

Restraining channel

PLAN

As the land drain is only to collect rainwater (surface water), there is no need to collect it for treatment. We simply need to collect it to prevent nuisance or to help improve the stability of the ground, and once collected there are therefore a number of options for disposal. Some of these options are illustrated in Figure 2.19.

The catchpit detail is effectively a break in the pipe to allow sediment to be collected before connecting the water to the true house drainage system. Without a catchpit a serious threat exists of sediment carried by the groundwater filling the house drainage pipes and disrupting the flow.

These solutions are permanent methods of water control, but we may wish to lower the groundwater level or control water in some other way on a temporary rather than a permanent basis. Water levels in the ground can be substantially reduced on a temporary basis by using the technique of dewatering. Here we surround the excavation area with well extraction points, which consist of perforated steel tubes which are connected with flexible hoses to a larger pipe that encircles the excavation site. This larger ring pipe is then connected to a pump. To sink the well pipes (generally referred to as *well points*) we force water around the horizontal ring pipe and down the well point. By doing this we jet a cavity by pressurised water, allowing the pipe to be sunk to the desired level. Once this is achieved the well point is surrounded by granular material to act as a filter when the pump is reversed and the water is extracted from the ground. Figure 2.20 shows how the water level in the ground may be influenced by the technique.

If we did not wish to lower the water table level but simply protect the excavation site from water we could use some temporary steel sheet piling as an intercepting barrier to the water. When such a barrier is formed into a complete enclosure surrounding the area of the site, we call it a *cofferdam* (Figure 2.21).

REVIEW TASKS

- Which part of the Building Regulations extensively applies to preparatory works for a new building site?

- Define the abbreviations SL and FL, which are associated with the excavation works needed for the formation of a ground floor.

- Visit the companion website at www.palgrave.com/engineering/riley1 to view sample outline answers to the review tasks.

3

The building process

AIMS

After studying this chapter you should be able to:

- Appreciate the nature of alternative approaches to the building process
- Understand the reasons for evolution of these various approaches
- Evaluate the benefits and disadvantages of these different approaches to construction
- Have an awareness of the sequence of operations involved in constructing dwellings
- Understand the principles of modern methods of construction
- Identify the key elements of expenditure on buildings
- Appreciate the link between house construction and sustainability

This chapter contains the following sections:

INFO POINT

- *Code for sustainable homes – setting the standard in sustainability for new homes* (2008)
- Sustainable Communities Act (2007)
- BRE Information Paper 3/07 Modern methods of construction in housing
- *Housing statistics* (2007), Department for communities and local government
- *Building a Greener Future: policy statement* (2007), Department for communities and local government
- *Building a Low Carbon Economy – the UK's contribution to tackling climate change* (2008), TSO Committee on Climate Change
- DEFRA (2001) *Limiting thermal bridging and air leakage. Robust construction details for dwellings and similar buildings*
- NF9 *Zero Carbon: What does it mean to home owners and house builders?* (2008), NHBC Foundation
- NF41 *Low and Zero Carbon Homes – understanding the performance challenge* (2012), NHBC Foundation
- *Introduction to PassivHaus; A Guide for UK Application* (2008), Building Research Establishment, Garston, Watford

3.1 | Methods of building

Introduction

- After studying this section you should have developed an understanding of the ways in which the building process has evolved.
- You should be familiar with the varying approaches to building, from the traditional to the industrialised.
- You should be able to identify the key aspects of each approach.

Overview

Buildings are formed by the assembly of large numbers of individual elements and components, of varying size and complexity. Historically the production and assembly of these components would have taken place on-site. As discussed in Chapter 1, vernacular architecture has developed as a result of the ability to fabricate components from locally available building materials. This was dictated by limitations in the ability to transport materials and fabricated components over even modest distances. With advances in transport networks and technology, notably during the Industrial Revolution, it became possible to transport materials and components over large distances. These advances introduced not only the possibility of using non-local materials, but also of producing even sizable components away from the site. The possibility of mass producing components in a factory environment initiated a change in approach to the whole building process, with the beginnings of industrialised building.

Traditional building

As discussed earlier, building form has historically been dictated by the availability of building materials, the local climate and the lifestyle of the population. The need of nomadic peoples to re-site their dwellings regularly imposes very different constraints on building form from those of people with a more static lifestyle. These forms of vernacular architecture could be considered to be the results of truly traditional building production. The nature of *traditional building* as it exists at present derives from this type of construction and the principles involved in it.

The use of traditional, labour-intensive, building crafts in the production of buildings is now restricted to the building of individually designed, 'one-off' structures or to the area of specialised building refurbishment. The cost implications of such a building method are considerable, with highly trained craftsmen fabricating components on-site, although a number of prefabricated components would inevitably be used in all but very specialised cases. The nature of this method, in which most parts of the buildings are formed from a number of small parts, made to fit on-site, is inherently slow. Hence its use for large buildings, or large numbers of buildings, in today's economic climate where time and cost are of the essence, is impractical.

There are, however, a number of advantages in the adoption of this method, in that it allows tremendous flexibility, with the potential to ensure that all parts can be adapted to ensure a good standard of fit, since all parts are 'made to fit'. This allows for flexibility both during the construction stage and throughout the life of the building. Elements of this mode of construction are still to be found in some modern building, such as house building using traditional rafter and purlin roofs. In this case the roof structure is fabricated by carpenters, on-site, from straight lengths of timber. Even this type of traditional building has become rare, with increased levels of off-site fabrication.

One of the main disadvantages of traditional building using on-site fabrication and the formation of components *in situ* is the difficulty of manufacturing components in a hostile location. The vagaries of the climate and conditions which prevail on-site restrict the ability to work to fine tolerances and to fabricate elements with consistency. Although building is essentially a manufacturing process, the conditions in which it is undertaken are far removed from those in a cosseted factory environment. The disadvantages of cost, time and uniformity of standards of quality, together with difficulties in obtaining suitably skilled labour, have resulted in the adoption of what has become known as 'conventional' or 'post-traditional' building.

Post-traditional building

The evolutionary nature of the building process, with the periodic introduction of new materials and new techniques, has ensured that even traditional building has developed and progressed significantly in the past. Examples of such developments are the introduction of ordinary Portland cement, allowing the production of large, complex building sections by casting of concrete, and the development of reinforcing techniques using steel, allowing very strong sections to be produced. Such advances in building technology, allied with the need to minimise construction time and cost, resulted in the adoption of the post-traditional method of building. It must be noted also that this would not have been possible without the advent of developments in transport mechanisms. These have allowed non-local materials to be transported to the site and components to be produced some distance away for transportation to the site.

This form of building is a combination of traditional, labour-intensive, craft-based methods of construction, with newer techniques, utilising modern plant and materials. The use of mechanical plant is one area where post-traditional building differs greatly from traditional building. Post-traditional construction is often adopted for the erection of buildings on a large scale. In such instances, the craft-based techniques of traditional building, such as plastering and joinery, are not excluded from the construction process, but rather are aided by the use of mechanised plant. Machinery used for earth moving, lifting of elements and mixing of concrete and plaster is now essential on most building sites as a result of the magnitude of the operations being undertaken. Additionally, increased use is made of prefabricated components, manufactured in large numbers, in factory conditions. Hence an element of industrialisation was introduced into the construction process. The nature of the building design has not evolved to the extent of system building, where a series of standard parts may be used to produce the end product. Instead, the mass-produced components are placed and finished in a typically traditional way, using traditional crafts such as joinery and plastering. An example of this approach is the use of prefabricated trussed rafters for roof construction, in favour of the traditional site-fabricated rafter and purlin form of construction. It is now very unusual to find the on-site fabrication of components such as windows and doors, which can be produced more cheaply and to higher quality standards in a factory environment. Although the traditional building activities have, in essence, changed little, the scale of post-traditional building, together with the demands of cost and time effectiveness, have placed great importance on the active planning of building operations. It is this planning to ensure efficiency which is one of the main trade marks of post-traditional construction.

Rationalised and industrialised building

Rationalised building, as considered today, is the undertaking of the construction of buildings, adopting the organisational practices of manufacturing industries, as much as this is possible, with the previously described limitations of the construction industry. Such an approach to the construction process does not necessarily imply the adoption of industrialised building techniques, but is more usually based upon the organisation and planning of commonly used existing techniques. The key to the effective use of rationalised building is ensuring the continuity of all production involved in all stages of the overall construction process. This continuity relies upon the evolution of building designs to allow full integration of design and production at all stages. The aim of this process of building is to ensure cost-effective construction of often large and complex buildings, within given time parameters, while maintaining acceptable standards of quality. This requires the construction process to be as near as possible continuous, hence necessitating the efficient provision of all resources in the form of labour, plant, materials and information. This can, to some extent, be enhanced by the use of standardised, prefabricated components and effective use of mechanical plant, thus separating fabrication from assembly and reducing on-site labour costs. It will be seen that this approach is a logical evolution from post-traditional construction.

In the later part of the 20th century, notably in the 1960s, great demands were placed upon builders to construct buildings quickly and cheaply. To cope with these demands, systems of building were developed based upon the use of standard prefabricated components to be assembled on-site, largely removing the reliance upon traditional building techniques. Such systems attempted to introduce industrial assembly techniques to the building site. This approach is often termed 'system building'. Within this description two basic approaches exist: *open system* and *closed system* building.

Open systems of building, sometimes referred to as 'component building' utilise a variety of factory-produced standard components, often sourced from a variety of manufacturers, to create a building of the desired type. The construction of buildings adopting this approach makes little or no use of the traditional 'cut and fit' techniques of traditional and post-traditional building. An example of such an approach is the construction of lightweight industrial buildings, which are based upon designs that utilise a selection of mass-produced components that are not exclusive to the specific building. Hence flexibility of design is maintained.

In contrast, closed systems adopt an approach which utilises a dedicated series of components, specific to the individual building and not interchangeable with components made by other manufacturers. Such a method is beneficial to the speedy and efficient erection of buildings when aided by the use of large-scale mechanical plant. These systems, however, do not allow the adaptation of the design on- or off-site. This can be very restricting, particularly during the later life of the building, when changing user needs may require flexibility in the design. Such systems have also been subject, in the past, to many problems associated with on-site quality control and lack of durability of materials. These problems arose, in part, from a lack of familiarity of the workforce with the new building techniques, the use of untried materials and the need to construct quickly, thus encouraging the short-cutting of some site practices. The occurrence of such problems and the inherent lack of flexibility in buildings of this type have resulted in the general rejection of closed systems in favour of open systems.

The development of systems has led to the creation and use of a range of 'modern methods of construction' (MMC). These warrant separate consideration and are included in section 3.2.

Accuracy in building

In traditional construction, accuracy in building was to some extent ensured by the ability to make components to fit specific spaces. This ability to be flexible in the production of elements of the building has been removed to a great extent, as many components are fabricated away from the site. The industrialised manufacture of components dictates an increased emphasis on the accuracy of component sizes in order to ensure that mismatches are reduced to a minimum on-site.

The use of factory-produced building materials and components in even small-scale post-traditional construction has resulted in some standardisation of building component dimensions. For example, the widespread use of plasterboard which is manufactured in a range of sizes based on multiples or modules, of 600 mm, is made more efficient if minimal cutting of panels is required. Hence room sizes based on 600 mm modules are common, thus reducing fixing time and wastage of materials on-site. Such an approach is also evident in the manufacture of components which are designed to fit into openings in brick walls, such as windows, which are manufactured in a range of sizes which correspond to multiples of whole brick sizes. This is termed a modular approach to component size. Such an approach is based on simple logic, and is of increased importance when related to the construction of larger, more complex, buildings. When dealing with larger buildings, the size and number of components which must be assembled increases substantially. The degree of accuracy with which they are assembled must be adequate to ensure that the building is erected without undue difficulty and without the risk of compromising its performance. The modularisation of such buildings is immensely beneficial in maintaining an acceptable degree of building accuracy. This is sometimes effected by the use of 'dimensional coordination'. Dimensional coordination relies on the establishment of a notional three-dimensional grid, within which the building components are assembled. The grid allows for some variation of component size, providing a zone within which the maximum and minimum allowable sizes of a given component will fit. This variation in the sizes of elements of buildings is inevitable for a variety of reasons, including:

■ Some inaccuracy in manufacture is unavoidable because of the manufacturing technique. The production of a component in concrete is subject to size variation as a result of drying shrinkage following casting, together with great limitations in the manufacture of very accurate formwork.
■ The high cost of producing components with great accuracy may be substantial and considered unnecessary in a given situation; hence a degree of variation in size may be accepted.
■ The accuracy of location of the component, resulting from fixing variations, also has some effect.

Hence building components are not designed to fit exactly into a given space or position of a given dimension. Instead, allowances are made for jointing and component linking, taking into account possible variations. For these and other reasons, including the need to allow for thermal and moisture-induced variation in size following construction, a degree of allowable variation or '*tolerance*' in component size is an essential feature of modern building. The use of modular design and the allowance for tolerance, while maintaining acceptable accuracy, is of particular importance in the design and construction of system-built structures. In such buildings, ease, and consequently speed, of site assembly are of paramount importance. In such situations, the ease of assembly of the parts of the building depends upon a number of factors, including:

■ The degree of accuracy with which components have been manufactured
■ The degree of setting out accuracy on-site
■ The nature of construction and assembly methods on-site
■ The nature of jointing of components and the degree of tolerance which is acceptable in a given situation.

It must always be remembered that the production of components to very fine degrees of accuracy has a direct cost implication, which may not be justifiable in a given situation. For this reason, and those noted above, it is common to refer to component sizes of an acceptable range, rather than an exact dimension, and this may be given in the form of a nominal dimension and an acceptable degree of variation, such as nominal size 1200 ± 10 mm. The degree of accuracy required depends on the exact situation of the components, but may be of particular importance where structure and services interrelate, since services are generally engineered with finer degrees of tolerance than is the building fabric.

3.2 | Modern methods of construction

Introduction

■ After studying this section you should be familiar with the concepts underlying modern methods of construction (MMC).
■ You should understand the reasons for the adoption of MMC in house building.
■ You should appreciate the link between the technology associated with MMC and the programming and procurement of construction work.
■ You should be familiar with the generic approaches to MMC in house construction.

Overview

The term 'modern methods of construction' (MMC) does not yet have a definition that is accepted, universally, by the construction industry. However, it is generally accepted as a term used to refer to a range of construction methods that seek to introduce benefits in terms of production efficiency, quality and sustainability. The methods that are currently being introduced into the house building industry in the UK and elsewhere differ from 'traditional' approaches significantly. Many of the approaches considered as MMC involve prefabrication and off-site manufacture. There is an ongoing debate regarding the ways in which MMC could, and should, be classified. There are strong arguments for adopting a classification system based on the performance attributes of the building and the building process, such as quality and efficiency of production. However, the most commonly utilised classification system derives from that developed by the UK Housing Corporation. This system classifies MMC buildings by construction form and can be summarised as follows:

■ Off-site manufactured – Volumetric
■ Off-site manufactured – Panelised
■ Off-site manufactured – Hybrid

- Off-site manufactured – Sub-assemblies and components
- Non-off-site manufactured modern methods of construction.

These classifications will be used for the purposes of this text although many other terms that can be used to refer to MMC are in common usage. Commonly used terms include:

- Modular building
- Industrialised building
- System building
- Prefabrication.

In order to understand the benefits of MMC it is important to appreciate the context in which they have developed and to recognise the drivers behind their evolution. The following sections will deal with the reasons for the development of the various approaches, together with providing an overview of the different approaches.

Background to the development of MMC

In considering the reasons for the development of MMC in the context of this chapter it is important to note that the consideration of the topic is undertaken in the context of house building. The reasons for developing MMC for house construction share many drivers with other sectors of the construction industry. However, some aspects are very specific to the house building sector.

Supply and demand in the house market

At the time of writing this book (2013) the UK housing market is experiencing a significant shortage of affordable homes in certain areas of the country. The adoption of MMC forms of construction provides a mechanism by which the rate of construction of new homes, particularly in the public sector, can be accelerated and maintained. Since many of the houses that will be built in the public sector are required to adopt, as far as possible, innovative techniques in their construction, the use of MMC will become more widespread. Although this refers specifically to the UK the principles are transportable in the sense that high demand, requiring fast development and construction, can be responded to readily by the adoption of MMC. Such an approach ensures effective supply and affordability.

Skills shortage

In recent years the nature of the construction industry, and of those entering it, has changed. The industry has suffered from problems of participation in training, and the 'boom or bust' image of construction has discouraged entry to some. Although the industry has made significant efforts to increase participation in training and many initiatives have been developed, this has tended to benefit the individuals that are employed permanently within organisations. In addition, the increased use of contract labour means that many workers do not benefit from these training programmes. The problem is exacerbated by the increased levels of activity in the industry at all levels. Hence, the problem of skills shortages has become a major issue in terms of both construction operatives and supervisory staff. Clearly this has implications on quality and pace of building.

More recently there has been a migration towards partnering and other forms of procurement that positively encourage stable employment and participation in training. However, this alone can not address the skills shortage problem. The use of MMC techniques, where much of the process leading to the construction of the building is undertaken in a factory production environment, reduces the need for skills on-site. Repetitive manufacture of components and building sections in a controlled factory environment assists quality assurance and reduces the 'risks' associated with quality control on-site. Quality control is still extremely important on the construction site, however it can be controlled more readily if the extent of site-based work is reduced and simplified.

Quality enhancement

The previous section relating to skills shortages made reference to the issue of quality assurance. The reputation of the construction industry in terms of its ability to maintain quality is mixed. Over the years much has been written about declining quality although the true situation is difficult to measure effectively. Without question there have been periods during which the industry has suffered from visible quality assurance problems; the aftermath of the 1950s and 1960s system building programme is a good example. In reality the issue of quality that is currently the concern for the industry hinges on enhancement rather than assurance.

The industry has sought to improve quality through advances in technology and training, while at the same time Government agenda for improved productivity, sustainability and 'right first time' principles have changed the context of the construction sector. Added to this, the increases in customer expectation, particularly in the housing sector, have forced a continuous programme of quality enhancement. The vagaries of the construction site environment and the fluidity of the contract labour market result in limitations upon the ability to enhance and maintain quality using traditional construction processes. Hence, the use of MMC using controlled factory manufacture, with site processes effectively limited to assembly of pre-manufactured buildings, provides an effective solution to the quality issue.

Developments in Building Regulations

The Building Regulations and their evolution were discussed extensively in Chapter 2 and it is not the intention to repeat that material here. However, the evolution of the Building Regulations and other similar frameworks outside England and Wales has had a significant impact upon the potential for the adoption of MMC in house construction. The requirements for threshold performance levels in terms of thermal and acoustic insulation and overall energy conservation demand a predictable and assessable construction approach. The potential for performance assessment of building post-construction introduces a much greater drive for repeatable, reliable construction detailing than was previously necessary. Clearly the use of factory manufacturing techniques lends itself to achieving reliability far more than does traditional site-based techniques.

Sustainability and environmental performance

It is generally accepted that around 10 per cent of all building materials that are ordered and delivered to site are unused and go to waste. The implications of such a wasteful construction process in terms of economics of production and sustainability are great. The increased focus on the environmental implications of construction in terms of process and whole-life performance places great emphasis upon the sustainability of building and approaches to building. The adoption of MMC techniques using factory manufacture and efficient materials use and supply chain management provide for a much less wasteful construction process. In addition, the quality control of MMC can result in higher levels of air-tightness, thus increasing thermal performance of the building in use. A further environmental benefit of factory production is that it removes much of the local environmental disturbance from the immediate site environs. Hence, sites adopting MMC will have lower levels of noise, dust and general nuisance than equivalent 'traditional' sites.

> **REVIEW TASKS**
> - Consider a typical house development that you are familiar with. How might the use of MMC affect the design, production and costs of the development?
>
> - Visit the companion website at www.palgrave.com/engineering/riley1 to view sample outline answers to the review tasks.

Procurement and programming of projects using MMC

The implications of MMC that utilise off-site manufacture of significant elements of the building affect the design, procurement and programming of projects. Unlike traditional projects, the pre-construction phase will encompass major activity in the creation of elements of the building.

The procurement and construction programme associated with traditional forms of building is essentially a linear sequence of activities. The completion of earlier stages is generally a prerequisite for the commencement of subsequent activities. A simplified sequence of activity is illustrated in Figure 3.1.

This model of procurement and programming restricts the construction activity to the period following detailed design and commencement on-site. If MMC techniques are adopted, much greater flexibility in the sequencing can be introduced since sections of the building superstructure and fit-out can be manufactured at the same time as substructure work is taking place on-site. The effect of this is to reduce the time required on-site for the construction phase. The greater the level of manufacture off-site, the lower the requirement should be for time on-site. However,

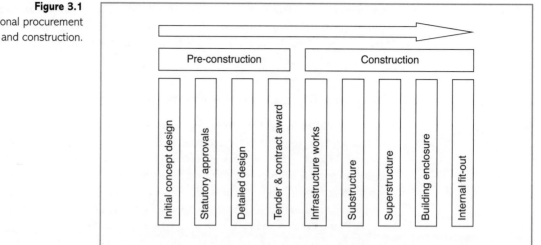

Figure 3.1
Traditional procurement
and construction.

increasing levels of pre-manufacture of building sections place much greater emphasis on the pre-construction phase in terms of programming. This is important in respect of the final agreement of the detailed design. Once the order has been placed for the manufacture of the panels, frames or volumetric units there is a requirement for 'design freeze'. The ability to amend the design after this point is extremely limited and will come with significant cost implications. Issues that need to be considered when procuring MMC buildings include:

- Design freeze: timing and extent
- Delivery timetable for units (storage on-site is undesirable but manufacturers will not want to store at the factory either)
- Responsibility for measuring on-site and agreement of tolerances
- Sanctions for late delivery and quality failures
- Protocols for accepting completed units and quality checking
- Defects liability periods and scope
- Manufacturer's role and responsibility during erection/assembly of units on-site.

Typical programmes of activity for MMC alternatives are illustrated in Figure 3.2.

Volumetric construction

Volumetric construction is sometimes referred to as 'modular construction'. This is not to be confused with the commonly used reference to temporary, portable buildings, such as site huts and so on, as 'modular buildings'. Modern volumetric construction is a well-developed and sophisticated process in which three-dimensional units or blocks are assembled and fitted out in the factory. These are then transported to site where they are assembled or 'stacked, on to pre-prepared foundations to provide the completed building.

The use of this technology is particularly suited to house construction as the resultant building is essentially cellular in form. Hence, there is limited flexibility in

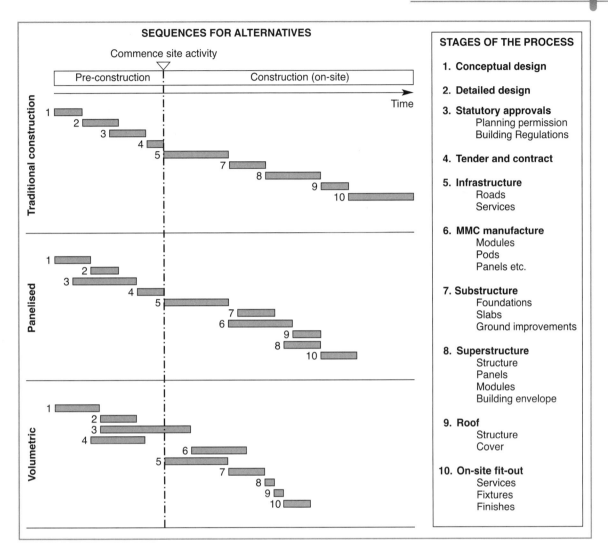

SEQUENCES FOR ALTERNATIVES

STAGES OF THE PROCESS

1. **Conceptual design**

2. **Detailed design**

3. **Statutory approvals**
   Planning permission
   Building Regulations

4. **Tender and contract**

5. **Infrastructure**
   Roads
   Services

6. **MMC manufacture**
   Modules
   Pods
   Panels etc.

7. **Substructure**
   Foundations
   Slabs
   Ground improvements

8. **Superstructure**
   Structure
   Panels
   Modules
   Building envelope

9. **Roof**
   Structure
   Cover

10. **On-site fit-out**
    Services
    Fixtures
    Finishes

Figure 3.2
Some activity
programmes for MMC.

the arrangement of the building during its design life. Many materials are used to create the modules or volumetric units, including timber, lightweight, cold-rolled steel and concrete. One of the most visible, and commonly utilised, forms of volumetric construction is the provision of bathroom 'pods' within existing or new buildings such as hotels, student accommodation and so on. The most effective use of volumetric construction is in buildings which have large numbers of identical room units, where repetition of the units can be achieved. Hence, their use is increasing in the development of flats, student residences and hotels. In house construction a completed dwelling will, typically, comprise four modular units enveloped with a pitched roof and an overcladding of traditional appearance such as brick or render panels.

Panelised construction

Panelised construction relies on the use of pre-manufactured flat panels that are assembled to create the building on-site. The panels are manufactured using a variety of materials, although timber and lightweight steel sections are the most common forms for housing. Chapter 7 (sections 7.3 and 7.4) deals with timber and steel frame construction and considers the technology associated with those forms in detail. The more advanced systems adopt panelised construction for intermediate floors and roofs as well as the wall units that are now commonly seen in the majority of timber and steel frame house developments.

Panelised construction (Figure 3.3) can be effected using a number of different approaches or types of panel which may be loadbearing or non-loadbearing depending on their position and function within the overall structure. The most common forms used in house construction are as follows:

Figure 3.3
Panelised construction options.

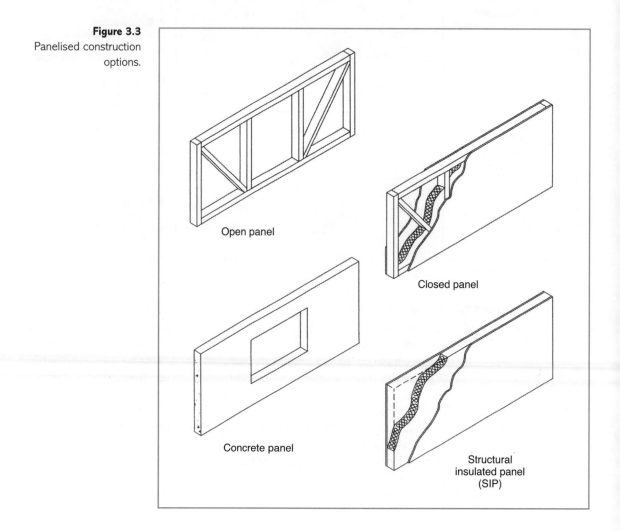

Open panel

The skeleton frames, typically in timber or lightweight steel, of these panels are manufactured in the factory. They are then transported to site where the insulation, vapour barriers, linings and claddings are fitted along with the services and components such as windows.

Closed panel

These panels are fully assembled and enclosed in the factory. They are likely to comprise the same components as described above for open panels but the entire assembly and sealing of the panels are based on off-site manufacture.

Concrete panel

Although rarely used in low-rise housing, the option exists for the use of precast reinforced concrete panels. These often provide a structural wall for the building and can incorporate windows, services, insulation and externally applied cladding materials such as brick 'slips' to provide a complete walling assembly. These panels are sometimes referred to as 'cross-wall' systems.

Structural insulated panel (SIP)

Unlike open and closed panel forms these panels do not have a skeleton structure to provide structural performance. Instead they comprise a central core of rigid insulation material that has external cladding and internal lining panels bonded to each side. The resulting, rigid panel can be used for forming wall and roof sections of the building.

Sub-assemblies and components

The use of prefabricated components within the traditional construction process is now widely accepted. The use of trussed rafter units for roofs could be taken as an example of this. However, as MMC systems have developed, the use of larger components or building elements that have been manufactured off-site has become more sophisticated and more widespread. The following components/elements may be found in modern MMC house construction:

- *Modular/prefabricated foundation systems*: See the Overview in section 4.2 of the following chapter for further consideration of these. Foundation systems based on the use of precast reinforced concrete beams supported by piles or piers allow fast assembly of foundations for dwellings.
- *Floor and roof cassettes*: Prefabricated panelised units for the floor and roof assemblies can be used to assemble floors and roof sections with reduced labour input, faster completion times and, importantly for roofs, accelerated weather-tightness of the building envelope.

■ *Roof segments*: Entire sections of pitched roofs can be factory assembled or assembled at ground level on-site. These are then craned into position to create the finished roof structure and fabric. The use of these technologies reduces the health & safety risks associated with working at height.

■ *Wiring 'looms'*: Adopting technologies that have been long-established in other industries, such as car manufacturing, the speed and quality of electrical installations can be enhanced using pre-manufactured 'plug-in' wiring looms. Cables systems are assembled in the factory to allow easy installation and the locations of sockets, lighting points and so on are terminated with plugs so that the finishing elements can simply be plugged in to complete the installation.

Non-off-site methods of MMC

Most of the techniques used for site-based MMC are applicable to industrial and commercial building rather than to low-rise house construction. As such they will not be considered in detail here.

REVIEW TASKS

■ Consider the examples of 'system building' described earlier in the chapter. How does the more recent development of MMC differ from these early attempts to industrialise house construction and how can the issues of quality control be managed better?

■ Visit the companion website at www.palgrave.com/engineering/riley1 to view sample outline answers to the review tasks.

REFLECTIVE SUMMARY

■ 'Modern Methods of Construction' (MMC) refers to construction methods that benefit production efficiency, quality and sustainability.
■ MMC involves prefabrication and off-site manufacture.
■ Commonly, forms of MMC for house construction include:
  – Volumetric
  – Panelised
  – Hybrid
  – Sub-assemblies and components.
■ Reasons for developing MMC for house construction relate to:
  – Supply and demand in the house market
  – Skills shortage
  – Quality enhancement
  – Developments in Building Regulations
  – Sustainability and environmental performance.
■ MMC techniques allow flexibility in sequencing of site operations so reducing the time required for the on-site construction phase.
■ Pre-manufacture of building sections places emphasis on the pre-construction programming and requires 'design freeze'.

3.3 | Building sequence

Introduction

- After studying this section you should appreciate that the construction process for the completion of a building may be divided into stages or phases.
- You should have an appreciation of the likely events that occur in the formation of a house.
- You should be familiar with the Plan of Work established by the Royal Institute of British Architects as a logical division of the stages of the building process.

Overview

The sequence of building may be viewed from two perspectives: the stages or phases into which the various design procurement and building operations fall, or a more detailed examination of the sequence of operations involved in the construction of a specific residential unit. For the purpose of this section both of these views will be examined.

Building sequence – the phases or stages of construction

Some years ago, the Royal Institute of British Architects (RIBA) established its Plan of Work. This was to divide the entire building process into logical stages in order to consider what should be happening at each stage and what the responsibilities of the parties associated with the process should be at each stage. The typical headings of the stages produced are:

- *Inception*   First idea of the building from client, definition of design brief and limitations
- *Feasibility*   Cost projections and assessment of possibilities
- *Outline proposals*   Sketches and conceptual design
- *Scheme design*   Scheme drawings and application for required planning approvals
- *Detailed design*   Working drawings and detailed assembly drawings
- *Production information* Specification of the work and workmanship requirements
- *Bill of quantities*   Material needs and quantification of elements
- *Tender action*   Examination of bids from contractors
- *Project planning*   Programming work and sequence of operations planned
- *Operations on-site*   Construction process
- *Completion*   Handover to client
- *Feedback*   Review of success, defects analysis and so on.

Once these headings have been established it is possible to produce a sheet of stages for each of the parties involved with the process, so sheets outlining the duties of the Architect and the Quantity Surveyor, for example, are produced. This not only helps to clarify the responsibilities of each person associated with the project, but also allows better planning and evaluation of the time likely to be associated with the process.

To have stages allocated to the design and building process is extremely helpful for the control mechanism of cost planning. When a project is first thought of (inception), some rough outlines are given to the construction economist (often the Quantity Surveyor) who is then asked to supply a likely contract figure with this bare minimum of information. The basis for this educated cost guesstimate is generally historic cost information stored from similar projects in the form of price per square metre of gross floor area (GFA). When a project is completed, the final contract sum is divided by the gross floor area of the building (that is, the floor area as measured to each floor inside the external walls and over any partitions or floor openings as if they were not there). The result is a cost per square metre.

This figure can be taken to make some arithmetical adjustments for the time delay between storing the cost information and the date proposed for the new project (index-based), and then the likely new building area is multiplied by the cost per m^2 to arrive at the first likely cost estimate (feasibility – is this within the client's cost range?).

As the design develops and specifications are firmed up, the detailed specification of the new building can be reviewed against those of historic building, making adjustments for quality and quantity as appropriate. In this way a running likely project cost is refined as the design develops, and it should be possible to ensure that the contract always falls within the budget of the client. If an overspend looks likely there would be an opportunity to reduce the specification, and hence costs, in a controlled and balanced way.

The way in which many construction contracts are arranged today (**procurement**) means that some of the details regarding the responsibilities of the professionals as listed on the Plan of Work may have changed, but the plan itself is still a useful breakdown.

Procurement is the term used to refer to the process of 'buying' a new building or a refurbishment project.

Building sequence – the stages in the construction of a house

Substructure is the term used to refer to works below or at ground level. **Superstructure** is the term that relates to the structure and fabric of the building above ground level. In reality, the shift between these elements occurs at the natural break-point of the DPC close to the ground floor.

When a house is constructed it is not possible to give an exact sequence of construction events, as one particular activity is not always commenced after another is completed. It is common for some activities to be undertaken at the same time. For example, the electrical installation could be undertaken while work is ongoing to the plumbing. External landscaping and planting could be happening at the same time as a variety of other tasks.

However, it is possible to provide a broad appreciation of the typical overall activity sequence for the construction of a house. For convenience this is often broken down into works 'up to DPC' and works 'above DPC'. This is a distinction between **substructure** works and **superstructure** works.

Up to and including external wall damp-proof course (DPC)

Setting out establishing the position and layout of the building on the site

When setting out the building, one of the key markers is the frontage line, and from this the layout of the rest of the building develops using lines projected at right angles to show the width of the property. A theodolite, builder's square or Pythagoras' triangle may be used to create the right angle (Figure 3.4).

Particularly when using strip footing foundations, the position of the trenches is marked for the excavator using profile boards (Figure 3.5). String can be stretched between these boards to represent the position of the trench and also the position of the wall.

Excavation of topsoil down to formation level (point of commencing floor) and foundation trenches (assuming strip footings)

compaction of trench bottoms

Concrete in strip foundations

Services route and position incoming services (and occasionally outgoing ground floor WCs), including ducts through walls to inside floor level as needed

install below-ground drainage system

form trench(es) for incoming water, electricity, gas and telecommunications

Figure 3.4
Setting out a right angle.

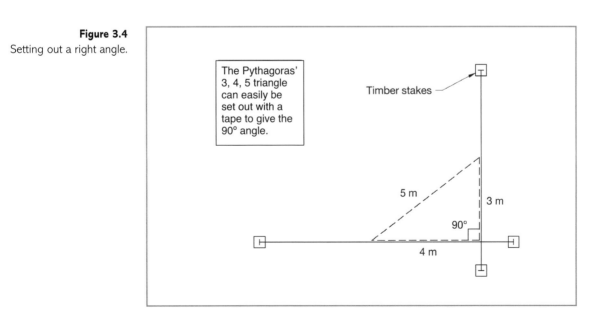

The Pythagoras' 3, 4, 5 triangle can easily be set out with a tape to give the 90° angle.

Timber stakes

5 m

3 m

90°

4 m

Figure 3.5
The use of profile boards
to mark trench positions.

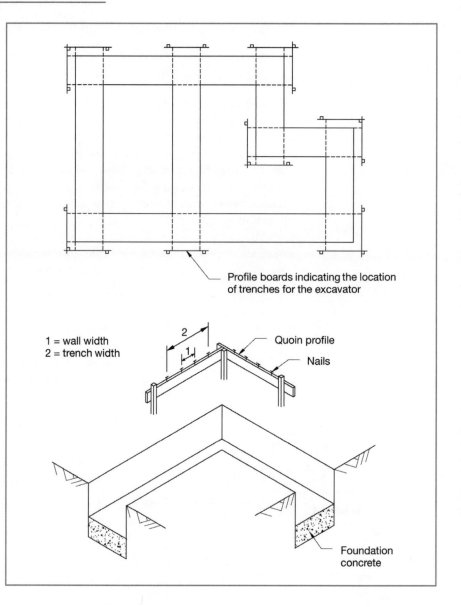

Profile boards indicating the location
of trenches for the excavator

1 = wall width
2 = trench width

Quoin profile

Nails

Foundation
concrete

Brickwork/blockwork	common bricks, blocks or trench block in foundation trenches
	facing bricks to outer skin up to DPC (three courses minimum is usual)
Filling	stone hardcore to floor slab and to inner part of foundation trenches
	sand blinding to hardcore surface
	the cavity to external ground level with mortar

Membranes	damp-proof membrane (DPM) (polythene sheeting, such as Visqueen) turned over inner skin of external wall and any internal walls which have foundations
Insulation	possible rigid board insulation for floor slab
Concrete	to floor slab (assuming solid floor slab)

Above external wall DPC

Brickwork/blockwork	construct external wall blockwork, place insulation and attach holding ties, lay brickwork
	form openings for windows and doors, including building in lintels, cavity tie/weep-hole arrangements and closing cavity
	construct brickwork/blockwork to internal partitions, all to first floor level
Woodwork	fix first floor joists
	fix staircase
Brickwork/blockwork	continue with external walls and window openings
Roofing	fix wall plate above inner external wall skin
	locate and fix trussed rafter roof structure
	fix sarking felt and tile roof and flashings
Partitions	fix the first floor partition carcass (assuming that all first floor partitions are non-loadbearing studded type)
Services	provide first fix for electrical (cables in)
	provide first fix for plumbing and heating (pipes in)
Windows and external doors	fix in position
Woodwork	apply floor boarding to first floor
Finishes	apply plasterboard and *in-situ* plaster finishes
	apply floor screed to ground floor (if not power floated)
Woodwork	fix skirting boards

Services	second fix electrical (lights, sockets and switches)
	second fix plumbing (sanitary appliances and connections)
	second fix heating (equipment and radiators)
	electricity company connection of supply
	commission services on completion including test of below-ground drainage
Fittings	install cupboards
	fit kitchen
Painting	externally and internally
External works	landscaping, grass laying, planting, boundary fencing/walls and paths

The completion of activities such as are listed above may be charted and a time allocated to them in order to produce a programme for the works. This would allow examination of progress to ensure that completion on time is achieved. This may not be such an issue for a single property, but if the project was a residential development of sixty units the establishment of a programme would be extremely useful.

REVIEW TASKS

■ Explain the purpose of the RIBA Plan of Work.

■ What do the first three stages of the RIBA plan involve?

■ Consider the various stages identified in the construction of a house and try to allocate times for each task, and how long it would take to build houses with reference to the various tasks identified.

■ Consider which items must be completed before commencing other items, and which can take place together. Try to generate a sequence of events and timescale for your work.

■ Visit the companion website at www.palgrave.com/engineering/riley1 to view sample outline answers to the review tasks.

PART 1

3.4 | Expenditure on building

Introduction

- After studying this section you should have gained an appreciation of the link between expenditure and time while constructing a property.
- You should appreciate that an *S curve* profile is the way in which we may represent expenditure commitments during the construction process.

Overview

The nature of the building industry is such that the vast majority of houses are built for profit by developers. In a relatively small number of cases the construction of dwellings is commissioned by individuals who are not driven by the Developer's equation that was discussed earlier. However, most house building is speculative, and the **Developer's equation** is the key to the viability of the project; hence the control of building costs is essential.

The **Developer's equation** was introduced in Chapter 2 (section 2.1). It is a simple relationship between land cost, income, outlay and profit.

The expenditure profile of any construction contract is directly related to the sequence of building operations. In the case of most housing developments, the 'contract' will generally encompass the entire development, which will include several individual dwellings. There are a number of implications arising from this kind of project, relating to the fact that the completion of individual units or phases of the contract is achieved prior to completion of the entire project. It is common to see housing developments in which differing stages of the construction process are visible at one time. Indeed, it is normal for parts of the development to be occupied by purchasers while other parts are in the early stages of construction. The reason for this is related to the Developer's equation and the need to generate income on the project as well as controlling expenditure.

Building costs

The total cost of a building project is made up of a number of individual elements, in addition to the cost of the materials, labour and plant used for the building itself. Aspects of setting up the site, providing insurance and so on are all elements that have a cost implication. These are often referred to as *preliminary items* and are normally included within the overall costings on the basis of a percentage addition to each costed item or as specific items identified within the project costings. It is not the purpose of this book to consider the nature of construction contracts and tendering; hence, we will not consider the basis of pricing in detail. However, it is important to understand the nature of the total project cost and the individual elements that it comprises. This was considered to some extent when the Developer's equation was introduced in Chapter 2, although the elements were not set out in detail. Since every construction project is individual, it is not possible to set out a definitive list of items of expenditure. Table 3.1 sets out the typical list of items included within the total cost of a building project from the point at which work commences on-site, but this should not be considered as definitive.

Table 3.1 Items included in cost of building project.

Item	Description	Phasing
Statutory and other fees	Costs associated with inspection of the works in progress, insurances and guarantees such as NHBC	Various stages within the project
Insurances	Public and employer's liability insurance, together with insurances for equipment, plant and buildings paid by the contractor and subcontractors	Normally paid at the outset of the project
Utilities connections	Costs of connecting to gas, electric, water, drainage and other utility services	During the early stages of the project
Temporary works	Protective hoardings, scaffolding, fencing, temporary supplies of power, water and so on	At the early stages of the project
Accommodation and facilities	Site accommodation for offices, canteen facilities, toilets, secure storage of plant and equipment and so on	At the outset of the project
Plant hire	Hire of specific items of plant and equipment, such as excavators and hoists	Throughout the project
Infrastructure	Roads, access and services installation	At the early stages of the project
Labour	Costs of skilled and unskilled labour to undertake the work, including supervision costs	Throughout the project
Plant	Costs of purchase, hire or depreciation of existing plant involved in the works	Throughout the project
Materials	Costs of materials and components used in the building process	Throughout the project
Making good	Costs associated with removal of temporary works and making good damage to any areas affected by the works outside the boundaries of the site	At the end of the project

Expenditure profile

As previously stated, the majority of house building projects include several dwellings. However, in order to understand the totality of the project we must first consider the expenditure profile relating to an individual house.

The preceding section considered the process of building and set out in some detail the typical sequence of operations involved in the construction of a dwelling. Each of these operations has a cost attributed to it that will be made up of three components: materials, plant and labour. The materials are the raw materials (such as timber, sand and cement) and the manufactured components (such as windows, lintels and roof trusses) that are required to facilitate the construction of the building. 'Plant' is the term used for mechanical tools and vehicles, such as lifting equipment and dumper trucks used to assist the construction process. 'Labour' is the human resource in the form of skilled and unskilled tradespeople taking part in the construction process. Each of these has a direct cost that can be calculated for any element of the project.

By applying these costs to each activity within the construction sequence we are able to build up an expenditure profile that illustrates the cost of the project at any given point. When expressed in terms of a cumulative cost profile, that is, the total cost of all elements up to any given point, this produces a curve which is referred to as a 'lazy S' (Figure 3.6).

The profile of the curve arises because expenditure accelerates at the beginning, achieves a consistent level in the middle and decelerates at the end of the project. There is a period at the beginning of the project during which the site is established and preparation works are underway. This results in modest outlay initially, before expenditure on materials and assembly begins. Once the preparatory period has been completed, expenditure on plant, labour and materials remains fairly constant

Figure 3.6
Plotted project expenditure – the S curve.

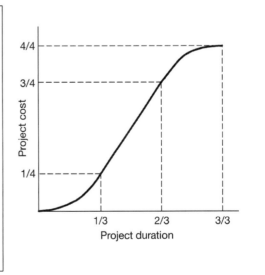

The typical 'lazy S' curve can be plotted for most construction projects and is used for forecasting cash flow and monitoring progress. In most cases the first third of the project accounts for around one quarter of the expenditure; the final third accounts for around one quarter and the remainder is spent uniformly during the middle third of the project. Note that the expenditure accelerates at the start of the project and decelerates at the end resulting in the distinctive S curved profile.

until the later stages. At this point, items of finishing and tidying are undertaken. These take time, but are not costly in terms of materials and so on. Although each project is individual, some generic models of expenditure have been developed. One of the most common is the quarter:third model. In this, the expenditure is anticipated on the basis that one-quarter of the cost will be attributable to the first third of the project period, one-quarter to the final third and the remainder spent uniformly in the period between.

Financing the building process

One of the factors that affects the profitability of any building project is the cost of financing the process. In most cases, the house builder will need to borrow money to fund the project before receiving payment for the dwellings. The cost of borrowing relatively large amounts of money can be significant and this must be taken into account by the builder as part of the Developer's equation. It would normally be the intention that such borrowings were of as short a duration as possible, to reduce the interest payments that must be made. This is one reason why building projects often have phased completion of dwellings. If some dwellings are available for early sale, the profit made can be used to reduce borrowings or to fund the later phases of the development directly. It is not uncommon to see large housing developments completed in increments, such that houses in one phase may be at foundation level while others have already been sold and occupied.

It is not the intention of this introductory text to consider in detail the funding and management of the construction process. However, these factors must always be borne in mind, since construction is a business and, like any other business, it is driven by profit.

REVIEW TASKS

■ What do we typically use to visually represent the expenditure outlay on a building as it develops on-site?

■ What would variations in this shape suggest?

■ Why is cash-flow important to developers in managing a successful project?

■ Visit the companion website at www.palgrave.com/engineering/riley1 to view sample outline answers to the review tasks.

3.5 | Sustainable house construction

Introduction

■ After studying this section you should be aware of the concepts surrounding sustainable communities.

- You should understand the reasons for the development of sustainable house construction.
- You should appreciate the key issues surrounding house construction and sustainability.
- You should be aware of the key technical issues affecting sustainability in house construction.
- You should be familiar with the concept of carbon neutral house construction.

Overview

Before considering the technology associated with the design, construction and occupancy of sustainable housing it is important to understand the broader context of sustainable development and sustainable communities. The principle of sustainable communities encompasses all the aspects that support the creation of homes and social environments that are environmentally responsible, that people want to live in, that encompass the entire social structure and that encourage positive interaction between all stakeholders.

Modern design and development of housing that supports these principles would be considered to be 'sustainable' in both technical and social contexts. This text is aimed at developing an understanding of the technology associated with sustainable house construction, although an understanding of the wider social, environmental and ideological issues is essential in appreciating the reasons for utilising such technologies. In general we would identify sustainable design as being associated with, among other things, incorporation of flexible design solutions that 'future-proof' the building, and technologies and design solutions that seek to reduce energy consumption such as the use of natural daylight and ventilation.

In the UK the Government has set a target to reduce carbon dioxide emissions by 80 per cent against 1990 levels. Since the housing stock of the UK is responsible for around 30 per cent of all carbon dioxide emissions, there is a natural focus on improving environmental performance in this area. Accordingly, it is the intention that all new homes in England will have to be zero carbon by 2016. In order to achieve this aspiration an environmentally sustainable approach must be taken to the design, construction and occupation of houses.

The introduction and implementation of a range of incentives and directives are intended to support radical change in the way in which we design and construct new homes. Implicit in the modern approach to house construction is a strong commitment to protecting and enhancing the environment, with particular focus on lowering carbon emissions.

Sustainability and house building

There is considerable scientific evidence showing that climate change is a serious and immediate issue. It is estimated that more than 25 per cent of the UK's carbon dioxide emissions result from energy used to heat, light and operate people's homes. It is clear that it has become essential that houses are built in a way that minimises the use of energy and emissions. In addition to direct energy consumption, the building and occupancy of houses have other environmental impacts due to water use, waste generation and the use of polluting materials. All of these impacts can be reduced by integrating sustainability principles into the house design and construction process.

Benefits of sustainable house construction

The house building sector of the construction industry is committed to improving the sustainability of its products. It is important to appreciate the benefits of such a shift in the approach to design, construction and occupancy to the various stakeholders that are involved. The benefits that can be identified could be considered within three broad groupings: general environmental benefits, benefits for developers and constructors, and benefits to occupiers or end-users.

General environmental benefits may be identified as follows:

- *Reduced greenhouse gas emissions*: With a reduction in greenhouse gas emissions to the environment will be a reduced threat from climate change.
- *Better adaptation to climate change*: With developing approaches to energy conservation, dealing with the effects of solar gains in summer and with increased standards of water efficiency, including better management of surface water run-off, the future housing stock should be better adapted to deal with the impacts of climate change which seem inevitable.
- *Reduced overall impact on the environment*: Inclusion of design measures that reduce the use of polluting materials and facilitate household recycling should ensure that future housing stock has fewer negative impacts overall on the environment.
- *Reducing the environmental 'footprint'*: By building more sustainable homes, and adopting responsible materials and supply chain management, the 'footprint' on the environment can be reduced.

Benefits for developers and constructors may include:

- *Improved comfort and satisfaction*: Homes built with sustainable principles embedded will enhance the comfort and satisfaction of end-users. This can provide benefit in terms of reducing complaints from purchasers and in allowing developers to differentiate themselves as 'environmentally responsible'.
- *Raised sustainability credentials*: The adoption of sustainability principles by housing providers can be used to demonstrate their sustainability credentials to the public, purchasers, investors and funding bodies.
- *Improved profitability*: The levels of wastage currently experienced within construction lead to lost profit for developers. Adopting more sustainable approaches to materials ordering and management should impact directly on profitability.

Benefits to occupiers and end-users may include:

- *Assisting choice*: The application of sustainability ratings will provide useful information for purchasers relating to the relative sustainability performance of different homes, assisting them in making informed choices.

- *Lower running costs*: Homes built with sustainable design features should have lower running costs resulting from greater energy and water efficiency.
- *Improved wellbeing*: Ideally, houses built using the sustainable principles outlined here should provide a more pleasant and healthier place to live. The use of increased levels of natural light and ventilation, for example, can improve the users' local, physical environment.

Framework for sustainable house construction

As part of its general approach to developing a sustainable environment the UK Government has initiated a series of measures intended to support the shift towards a genuinely sustainable construction industry. The Government is seeking to provide a set of measures intended to establish a rigorous framework to ensure environmentally sound practice for those involved in the house building industry in the UK. These include:

Building Regulations
The inclusion of stringent requirements for the conservation of fuel and power within the Building Regulations in England supports the drive for sustainable construction. Specifically, Part L of the Regulations places requirements upon designers and builders to deliver buildings with low levels of heat loss and with prescribed levels of 'air-tightness'.

Code for Sustainable Homes
This is a national standard for sustainable design and construction of new homes. The Code aims to increase the sustainability of new homes and provide information for purchasers. New residential developments receive a star rating from one to six, based on a structured assessment against the principles of the Code. Initially, use of the Code was voluntary, although some planning authorities did apply conditions specifying its use. From April 2008 it became mandatory to assess new homes against the Code to demonstrate 'carbon performance'.

Planning policy
Planning policy will be refined to deal with mechanisms for providing new homes, jobs and infrastructure needed by communities which tackle climate change and to set out policies for renewable energy.

Sustainable design principles
Assessment of the level of sustainability of house design and construction needs to take account of the dwelling as a complete package. Rather than considering the design, construction and occupation as separate issues, the 'whole home' approach seeks to recognise the totality of the dwelling's impact upon the environment in an integrated way. In order to do this the UK's *Code for Sustainable Homes* identifies a range of categories against which sustainability can be assessed. These include:

- Energy and carbon dioxide emissions
- Pollution

- Water efficiency
- Materials use
- Surface water run-off
- Waste management
- Health and wellbeing
- Management
- Ecology.

The drive towards sustainable design and construction is also reflected in the Building Regulations, which are the minimum standards required by law. Standards above the requirements of the Building Regulations are now being delivered by many environmentally aware developers who wish to differentiate themselves in a competitive marketplace. The *Code for Sustainable Homes* is a UK-specific initiative; however, it provides a useful framework that can be used to demonstrate the key principles of sustainable house building. These principles are internationally transportable and the UK Code is referred to here as an exemplar of the general issues.

The Code uses a rating system, which awards stars to dwellings to indicate the overall sustainability performance. Sustainability rating can be achieved from one

Table 3.2 Code requirements for achievement of flexibility within each sustainability category.

Category	Requirements for achievement of the Code (flexibility)
Energy/carbon dioxide emissions	Compulsory. Considered of greater importance and allowing least flexibility in achievement of the requirements of the Code. Minimum levels of performance are set for each star rating
Pollution	Flexible. No minimum standards set
Water efficiency	Compulsory. Considered of greater importance and allowing least flexibility in achievement of the requirements of the Code. Minimum levels of performance are set for each star rating
Materials use	Some flexibility. Minimum performance levels set for one star performance
Surface water run-off	Some flexibility. Minimum performance levels set for one star performance
Waste management	Some flexibility. Minimum performance levels set for one star performance
Health and wellbeing	Flexible. No minimum standards set
Management	Flexible. No minimum standards set
Ecology	Flexible. No minimum standards set

to six stars depending on achievement of the standards that make up the Code. One star is the entry level (although this is still above the minimum level of the Building Regulations) and six stars are the highest level achieved by exemplar sustainable development. All star ratings are calculated on a points basis across the nine categories identified above. There is flexibility in how the overall points score is achieved, but some performance aspects are compulsory. Table 3.2 summarises the requirements of the Code in terms of the flexibility within each category.

It is not appropriate to consider the Code in great detail within a text such as this. However, the issues that must be considered when allocating points within each of the nine categories may be identified as examples of the issues that should be considered in the sustainable design and construction process. These issues are valid in all areas where there is a wish to move towards sustainable construction and occupancy of dwellings. Although developed in the UK the principles are applicable globally. Elements that may be considered within the various categories and their implications for design, construction and occupation are summarised in Table 3.3.

Table 3.3 Elements to be considerd within each category of the Code, and their implications.

Issues considered	Implication for design, construction and occupation
Energy and carbon dioxide emissions	
Heat loss through building fabric	Thermal insulation, areas of windows relative to walls and levels of air-tightness of the building envelope must be considered in design and construction
Internal and external lighting	The use of energy-efficient light fittings and timers/automated switching systems should be considered and included
Drying space	Provision of dedicated drying space reduces the need for mechanical drying of clothes and so on, so decreasing environmental impact in use
Eco-labelled white goods	The use of eco-labelled white goods (cookers, fridges and so on) assists environmental performance of the dwelling in use
Low or zero carbon energy technologies	The inclusion of facilities for exploitation of renewable energy sources, such as solar panels, wind turbines and so on, can assist in reducing energy requirements. These elements are most effective when integrated into the initial design rather than grafting them on to traditional designs as 'extras'
Cycle storage	The provision of cycle storage within the design supports wider transport planning initiatives that support a national sustainability agenda
Home office	As more people benefit from flexibility in the location of their work, the facility for 'home working' can reduce the levels of energy required for commercial buildings, transport and so on

Pollution

Global warming potential of insulation material	The specification of insulation materials that do not result in the production of ozone-depleting gases should be ensured
Nitrous oxide emissions	Careful specification of heating and other systems is important to reduce nitrous oxide emissions to the atmosphere

Water efficiency

Internal potable water	Reduction in the levels of potable water should be ensured by the incorporation of efficient appliances, use of low flush cisterns and, if appropriate, use of grey water for some uses
External potable water	The collection of rainwater, using water butts for example, for some uses is effective in reducing the levels of external use of potable water

Materials use

Environmental impact of materials	Careful selection of materials for the construction of the major elements of the dwelling is important. Eco-labelling schemes can assist in ensuring appropriate selection. One of the key issues to be considered here is the extent of embodied energy within construction materials
Responsible sourcing of building elements	Specification and ordering of materials which are derived from environmentally responsible sources is important to ensure that sustainable practice is reflected throughout the supply chain

Surface water run-off

Reduction in surface water run-off	The control of surface water run-off has implications for the local environment and the extent to which sewers and drains are required to cope with excess water
Flood risk	The recent experience of large-scale flooding in many parts of the UK has reinforced the importance of consideration of the potential implications of localised development upon the delicate environmental balance

Waste management

Household recycling	Features incorporated into the design and the manner in which occupiers deal with the drive to separate waste to allow recycling combine in this area
Construction waste	The extent to which material wastage during the construction process can be minimised by the use of effective supply chain management techniques, better quantification of materials prior to ordering, good site practice and the adoption of dimensions for building elements that allow the use of modular materials without undue cutting on-site

Composting facilities	Features incorporated into the design and the manner in which occupiers deal with the drive to compost waste combine in this area

Health and wellbeing

Daylight	The incorporation of design features aimed at maximising the benefit of natural light within the dwelling will assist in providing a pleasant environment and in reducing energy consumption
Sound insulation	The passage of sound between dwellings can be a cause of nuisance and the design of adjacent units should avoid undue sound transmission through appropriate detailing. This must be supported by high levels of site quality control since even minor construction failings can lead to substantial sound transmission
Private space	Over the years, many models of social housing development have evolved with varying degrees of success. One of the issues that has become clear, however, is the importance to occupiers of private external space, often referred to as 'defensible space'
Lifetime homes	Increased focus has been placed on the lifelong performance of dwellings. One of the key aspects of ensuring high levels of lifelong performance is the concept of adaptability, allowing homes to be re-configured as the needs of the occupiers evolve through time. The use of flexible design and construction features assists in this

Management

Construction site impacts	Local environmental impact, materials wastage and construction quality can be controlled by the application of good site management
Security	The security of homes is seen as a significant element in the overall design, and features should be included to make dwellings inherently secure for their occupiers without compromising quality of life and personal freedom

Ecology

Ecological value of site	The ecological impact of the construction process must be minimised and/or controlled such that the design and construction processes manage the impact on eco-systems. Ideally, in a carefully managed development there may even be ecological benefit rather than damage
Ecological enhancement	
Protection of ecological features	
Building footprint	The environmental or ecological footprint of the dwelling must be benchmarked against good practice. This may be undertaken using an impact floor area ratio calculation model

Carbon dioxide emissions and dwellings

One of the key elements that will assist in achieving a sustainable, energy-efficient building is the control of heat loss from the building. The requirements for thermal insulation of the envelope are dealt with in the chapters dealing with individual structural elements such as walls, floors and roofs. However, the Building Regulations also set out requirements for entire building enclosure and include standards for the air-tightness of the building.

There is now a requirement to use a 'whole house' compliance method to assess the level of energy efficiency of dwellings. All new dwellings must demonstrate, at the planning stage, that an acceptable carbon dioxide emission rate will be achieved in the completed building. A Target Emission Rate (TER) must be calculated based on the building design information, and the actual Dwelling Emission Rate (DER), which is assessed after construction, must then be no worse than the TER. After completion of the building, an Energy Performance Certificate should be produced. This is based on the actual construction and should incorporate any changes to the performance of the building fabric or services between design and construction.

This energy assessment process requires benchmarking of actual building performance. In order to achieve the requirements of the Regulations, designers and constructors will need to deliver buildings with lower U (thermal transmission coefficient) values, greater levels of air-tightness, higher boiler efficiencies and the integration of renewable energy into the design. In order to achieve reduced emission levels it is necessary to limit heat loss through the building fabric and to minimise the extent of excessive solar gain and heat gains/losses from the elements of systems used for space heating, comfort cooling and hot water. Properly commissioned heating and hot water services are important in this context, as is the effective briefing of building occupiers to ensure that the dwelling and its services can be operated efficiently and effectively.

Air pressure testing of the completed dwelling must be carried out to ensure that there are adequate standards of air-tightness of the building fabric. The achievement of adequate standards of air-tightness must be dealt with at the design stage of the building. The remedial treatment of air leakage will be costly and inefficient. Hence, the use of 'robust details' to minimise air leakage by ensuring air barrier continuity and sealing elements at critical junction details is essential. Some of the main areas to suffer from air leakage are as follows:

- At connection points between walls and floors
- At positions of services entry
- Through porous masonry construction, including mortar joints and so on
- Around doors, windows and other openings in the building fabric.

In the UK, in order to encourage reduction in carbon dioxide (CO_2) emissions from dwellings, the concept of feed-in tariffs has been introduced. The principle of the feed-in tariff (FIT) is that home owners and occupiers are encouraged to install renewable and low carbon electricity-generating equipment. In return they benefit from financial incentives. The scheme is supported by legislation and installers and home owners can receive payments related to the amount of electricity generated from renewables or CO_2 reduced due to the provision of carbon reducing technologies.

In addition to the FIT initiative, the UK has introduced a variety of other incentive schemes to promote efficient energy use and reduction of carbon emissions. The 'Green Deal' initiative is aimed at reducing carbon dioxide emissions from existing homes by providing a mechanism for subsidised improvement measures for home owners.

At the time of writing (2013), the introduction of the Renewable Heat Initiative (RHI) is awaited. This aims to increase the proportion of domestic heating generated from renewable sources.

The UK Government has set ambitious targets for carbon reduction for new houses to be zero carbon by 2016. The exact definition of a 'zero carbon home' has been amended somewhat since it was first introduced. The term was originally intended to mean that 'no net CO_2 emissions would result from all energy used within a home during the course of a year'. This included 'regulated' and 'unregulated' energy use. Regulated energy use is that associated with heating, lighting, ventilation and so on, while unregulated use is that associated with domestic appliances, cooking and so forth. It is estimated that these energy uses can typically account for more than 3 tonnes of CO_2 emissions per year for a dwelling in the UK.

In 2011 the definition of 'zero carbon home' was amended within the zero carbon new homes policy to include only regulated uses. The Zero Carbon Hub thus provides advice and data on achieving the Government's targets for reducing carbon emissions. Figure 3.7 illustrates the progressive targets for carbon emission reduction in new homes up to 2016. In practice, the delivery of the 'zero carbon home' relies on a combination of energy-efficient building fabric, efficient building services and on- or off-site low and zero carbon heat and power solutions.

Figure 3.7
Progression of carbon reduction targets up to 2016 – zero carbon homes (excludes unregulated uses).

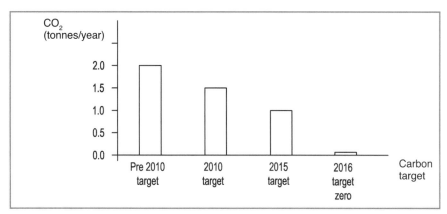

Air leakage and draught control

There is a natural tension in the design and construction of buildings between reducing the level of unwanted air leakage from the internal space and maintaining appropriate levels of ventilation where required. Modern dwellings are designed to be relatively tight in terms of air leakage, and pressure testing is used to measure unwanted air leakage – which can be a major problem in ensuring internal comfort for dwelling occupiers. Even where good levels of thermal insulation are achieved by the fabric, air leakage can allow heated air to escape, which is replaced by cold external air. This has two implications: firstly, the volume of air that must be heated is greater owing to the constant replacement of heated air with fresh, cold air; secondly, the movement of the cold air within the space gives rise to draughts and feelings of discomfort.

The potential exists during design and construction to reduce unwanted air leakage by adopting a series of simple procedures aimed at eradicating gaps and voids in the external enclosure. These may include, for example:

- Effective draught sealing for window and door assemblies
- Sealing minor gaps around openings
- Sealing gaps in walls around joist pockets and so on
- Forming timber floors with sheet materials rather than with separate floorboards
- Effective sealing around services entries and so on
- Sealing gaps between walls, floors and adjacent elements
- Ensuring tight-fitting access hatches are fitted into roof spaces and unheated voids.

However, the healthy occupation of dwellings relies on the maintaining of adequate ventilation to control moisture levels and to ensure the replenishment of clean air and the dilution of contaminants. This necessitates the provision of controlled ventilation at an adequate level. Hence, having made great efforts to reduce uncontrolled air leakage it may then become necessary to introduce controlled ventilation of the correct magnitude in the required spaces. This will generally be achieved by the utilisation of a combination of natural ventilation, possibly adopting a passive stack approach with heat recovery, and low-power mechanical ventilation that may be operated only when necessary. The control systems for these measures are important to ensure that they operate only when needed, and the incorporation of humidistats and thermostatic controls aids the process greatly.

In the UK, explicit air-leakage targets were introduced for the first time through the 2002 revisions to Part L of the Building Regulations. Differing approaches were accepted to achieve compliance including:

- Adherence to the DEFRA Robust Construction Details guidance (see DEFRA, 2001)
- Assessing actual air permeability through pressure-testing the building using the CIBSE approved method (air permeability not to exceed 10 $m^3/(h.m^2)$ at 50 Pa internal pressure).

Design **air permeability** is defined in the Building Regulations, Part L1A, 2006 as 'the value of air permeability that is selected by the designer for use in the calculation of the Dwelling Emission Rate'.

In practice the adoption of Robust Construction Details was far more widely used as the mechanism for achieving compliance and regulatory approval for new dwellings.

Subsequent revisions of the regulations (Approved Document L1A, 2006) have been made, requiring new dwellings to be constructed such that the external fabric should be of a reasonable quality of construction so that the **air permeability** is within 'reasonable limits'. A reasonable limit for the 'design air permeability' is stated as 10 $m^3/(h.m^2)$ at 50 Pa.

PassivHaus

PassivHaus or 'Passive House' originated in Germany in the 1990s and is an energy performance standard that focuses on the building fabric and services rather than the wider context of the dwelling that is adopted by the *Code for Sustainable Homes*. Although it is applicable to a wide variety of building types its adoption for dwellings, in particular, is increasing in the UK as a consequence of increasing concern about fuel costs, carbon reduction and the deliverability of broader standards such as the *Code for Sustainable Homes*. The intention of the PassivHaus standard is to build dwellings that have excellent thermal performance and air-tightness, aided by the effective use of mechanical ventilation. The aim is to minimise heating demand in the dwelling by adopting a structured approach to the design and detailing of the building fabric and services. The PassivHaus approach should result in the creation of dwellings that require minimal space heating and which facilitate effective use and recovery of introduced heat. The process of heat recovery and redistribution through the interior of the dwelling is made possible by the use of a Mechanical Ventilation and Heat Recovery (MVHR) unit. This approach also liberates benefits in terms of indoor air quality as a result of the ability of the MVHR unit to filter the air.

There are numerous approaches to the design and construction of energy-efficient homes, most of which rely on a combination of passive design features and active, efficient buildings services installations. It is argued that PassivHaus goes further than these by reducing space heating demand and primary energy consumption to an absolute minimum. In general the approach takes a 'whole house' view and relies on a series of key features aimed at delivering appropriate thermal comfort, while

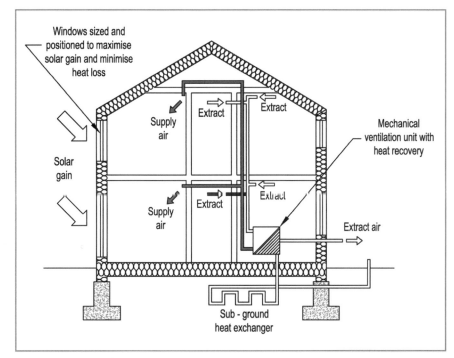

Figure 3.8
PassivHaus features.

PART 1

minimising the requirement for introduced heat. Some of the key features of PassivHaus design (see Figure 3.8) can be summarised as follows:

- High levels of insulation of building fabric
- Robust detailing to avoid thermal bridges
- Exploitation of passive solar gains
- Recovery from internal heat sources
- Very low air permeability
- Good indoor air quality (using MVHR).

Building substructure

4

Foundations

AIMS

After studying this chapter you should be able to:

- Identify the various elements that comprise the substructure of a dwelling
- Appreciate the interaction between the various elements and the implication of selection of specific design alternatives
- Relate the selection of individual options to building form, ground conditions and construction process
- Appreciate the criteria upon which selection of alternatives is based

This chapter contains the following sections:

4.1   Foundations
4.2   Functions of foundations and selection criteria
4.3   Types of foundation

INFO POINT

- Building Regulations Approved Document A, Structure (2004 including 2010 amendments)
- BS 1377: Methods of test for soils for civil engineering purposes (1990)
- BS 5837: Trees in relation to design, demolition and construction (2012)
- BS 5997: Guide to British Standard codes of practice for building services (1980)
- BS 6515: Specification for polyethylene damp-proof courses for masonry (1984)
- BS 8004: Code of practice for foundations (1986)
- BS 8103: Structural design of low-rise buildings. Code of practice for stability, site investigation, foundations, precast concrete floors and ground floor slabs for housing (1996)
- BS 8110: Structural use of concrete (1997)
- BS 8215: Code of practice for design and installation of damp-proof courses in masonry construction (1991)
- BS 8301: Code of practice for building drainage [no longer current, but cited in Building Regulations] (1985)
- BRE Digest 64: Soils and foundations (no date)
- *Efficient Design of Piled Foundations for Low-rise Housing – Design Guide* (2010), NHBC Foundation

4.1 | Foundations

Introduction

- After studying this section you should be able to distinguish between the various foundation options available for use in low-rise construction.
- You should have developed an appreciation of the functional requirements of foundations and the implications of soil type upon their selection.
- You should have developed a detailed understanding of the construction detail and sequence of operations associated with each of the forms in common use.
- You should be able to evaluate a variety of scenarios and make valid selections of foundation type based upon a detailed understanding of the issues involved.

Soils and their characteristics

Overview

The stability and integrity of any structure depend upon its ability to transfer loads to the ground which supports it. The function of foundations is to ensure the effective and safe transfer of such loadings, acting upon the supporting ground while preventing overstressing of the soil. The nature of the structure, its foundations and the soil onto which they bear dictate the ways in which this function is achieved. It is inevitable that in the period shortly after the construction of a building some consolidation of the soil takes place. Minor initial settlement is, therefore, to be expected; but more serious movement must be avoided, in particular that which is uneven and which may result in differential settlement, causing cracking and deformation of the building. The specific design of a foundation depends upon the structure, the way in which its loads are delivered to the foundation and the loadbearing characteristics of the soil type. It is logical, therefore, in considering the performance requirements and design of foundations, to first consider the various soil types and their individual properties.

Before attempting to study this section carry out the following exercise.

REVIEW TASKS

- On the basis of your general knowledge, try to identify the different types of soil that could form a base for the construction of low-rise buildings. Attempt to rank the various types in order of strength.

- Reconsider the exercise after studying this section.

- Visit the companion website at www.palgrave.com/engineering/riley1 to view sample outline answers to the review tasks.

Soil types

In practice, there is an infinite variety of soil compositions; however, these can be broadly categorised into five generic types in addition to solid rock, although these can be further subdivided into a large number of specific descriptions. Since the properties of soils depend upon the size of the particles from which they are comprised, the system of classification often used is based upon particle type and size. In addition to solid rock, the five categories are gravels, sands, silts, clays and peats (Table 4.1; Figure 4.1).

Table 4.1 Bearing capacity of different soil types.

	Rock	Gravel	Sand	Silt	Clay
Bearing capacity (kN/m²)	800+	Up to 600+	Up to 300+	Up to 75	75–300
Particle range (mm)	N/A	2+	0.06–2	0.002–0.06	<0.002

Figure 4.1
Typical particle size (upper limits) for various soil types.

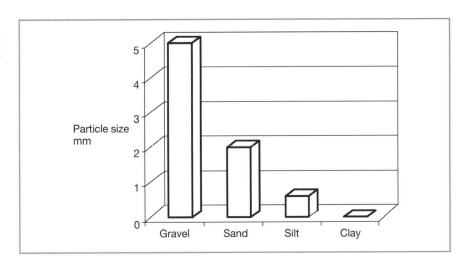

Identifying soil types

The procedures adopted for identifying the various soils on-site are set out in BS 8004, BS 8103 and BS 1377. It is clear that the size of particles making up the soil has an effect upon the strength and cohesion of the ground. These factors directly affect the ability of the ground to provide an appropriate supporting medium for foundations to low-rise buildings. As a general rule the level of cohesion of the soil increases as the size of particles decreases.

As discussed in section 2.2, concerning site investigation, the process of identifying soil types on-site is relatively simple. *Building Research Establishment Digest 64* sets out some of the salient characteristics of the various types, and attempts to identify some of the potential problems associated with each.

Rock

Solid rock provides a sound base upon which to build, as a result of its very high load-bearing characteristics. Typically the safe loadbearing capacity of sedimentary rock, such as sandstone and limestone, is approximately ten times that of a clay soil, with igneous rocks, such as granite, having capacities 20–30 times that of clays. These high loadbearing capabilities suggest that solid rock is an excellent base upon which to build. However, there are also great disadvantages resulting from the difficulty in excavating and levelling the substrata. This presents a considerable cost implication in building on such substrata. An additional problem, manifested in some sedimentary rocks, is that posed by the presence of inherent weak zones, created as a result of fault lines in the rock, together with the presence of slip planes between adjacent layers of differing composition. In such circumstances the use of ground securing anchors (Figure 4.2) is possible to alleviate the problem of instability.

Figure 4.2
Use of ground anchors to secure unstable rock layers.

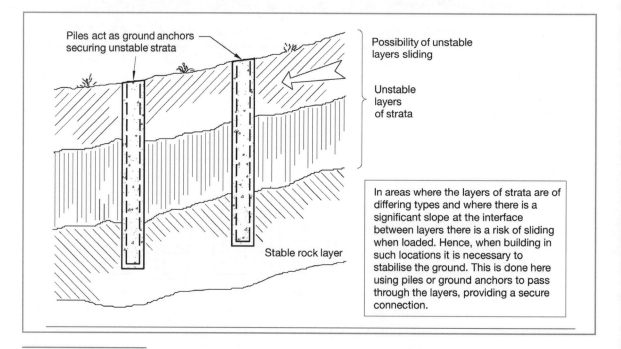

Piles act as ground anchors securing unstable strata

Possibility of unstable layers sliding

Unstable layers of strata

Stable rock layer

In areas where the layers of strata are of differing types and where there is a significant slope at the interface between layers there is a risk of sliding when loaded. Hence, when building in such locations it is necessary to stabilise the ground. This is done here using piles or ground anchors to pass through the layers, providing a secure connection.

Angle of repose is the term used to refer to the angle at which soil will settle naturally following excavation. Firm clay can be excavated without support and may leave an almost vertical side to the excavation; sand, in contrast, will tend to fall back into the excavation leaving a shallow angle at the edges.

Cohesive and cohesionless soils

From the point of view of the process of construction, the nature of the soil is potentially important for reasons other than its loadbearing capacity. When excavating for the placement of foundations the extent to which the soil is self-supporting, or conversely, the extent to which it requires temporary support, is also important. The need to provide temporary support to excavations has implications in terms of the speed of construction and the cost. In general terms, the more cohesive the soil, the greater its ability to support itself during the process of excavation. Thus the degree to which trench side support is necessary depends on the degree of cohesion of the soil as well as the proposed depth of the excavation. The **angle of repose** can be used as a guide.

Gravels and sands

Gravels and sands are often grouped together and considered under the broad heading of coarse-grained cohesionless soils. The particles forming these soil types, ranging in size from 0.06 mm upwards, show little or no cohesion; hence, if unrestrained, they tend to move independently of each other when loaded. This property can result in many problems if not addressed; however, if treated properly these forms of soil can provide an adequate building base. When initially loaded some consolidation is likely to occur, which may pose problems in relation to the connection of services and so on. The voids present between individual particles are relatively large, but represent a low proportion of the total volume. The presence of such voids results in the soils being permeable; hence water is not held in the ground for long periods.

Silts and clays

These types of soil are generally considered within the broad category of fine-grained cohesive soils. The small size of the particles making up these forms of soil, with a high proportion of small voids between particles, causes the tendency of clays and silts to display considerable variation in volume when subject to changing moisture content. This results in drying shrinkage in dry weather, and swelling and heave in wet weather (Figure 4.3). The nature of the soil is such that moisture is retained for

Figure 4.3
Ground moisture and foundation stability.

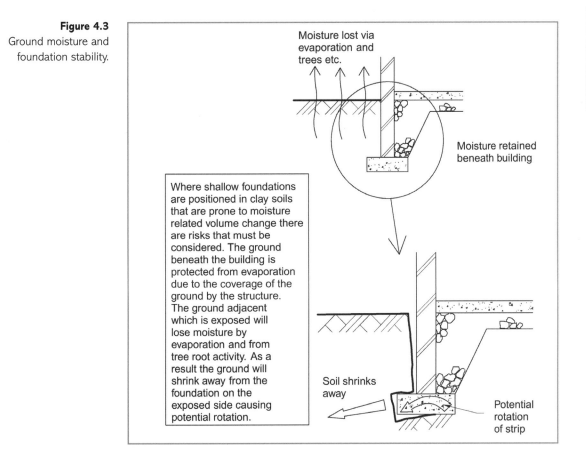

PART 2

considerable periods, because of the low permeability. Such soil types may also be compressible when loaded, especially at shallow depths. However, at suitable depths, below the region which is likely to be affected by moisture variation (normally 1 m or deeper) satisfactory bearing is normally found. The effects of trees, particularly those which have been felled recently, can be significant as they affect the amount of water which is removed from the ground. A further problem which is sometimes present in areas of clay soil is that posed by the aggressive actions of soluble sulphates in the soil upon ordinary Portland cement, which may necessitate the use of sulphate-resisting cements in some areas.

Peat and organic soils

Soils which contain large amounts of organic material are generally subject to considerable compression when loaded, and are subject to changing volume resulting from the decay of the organic matter. Hence they are generally unsuitable for building. For these reasons top soil is removed prior to building, and excavation will normally take place to remove soil with a considerable quantity of organic matter present.

Table 4.2 shows the **typical safe bearing capacity** of various soil types.

Trees and their effects

As mentioned previously, the effects of the presence of trees in close proximity to the foundations of the building can be damaging to the structure of the building

The range in bearing capacity between soil types is obviously considerable. However, it must also be noted that ranges within soils of a given type can be great; hence the term '**typical safe bearing capacity**' is used.

Note: this is a measure of pressure that the soil can withstand.

Pressure = Force/Area

The area of contact with the ground can be varied by appropriate foundation design.

Table 4.2   Typical safe bearing capacity of various ground conditions/soil types.

Soil type		Typical safe bearing capacity (kN/m²)
Rock		
Igneous		10 000
Limestone/strong sandstone		4 000
Slate		3 000
Shale		2 000
Cohesionless soils		
Gravel/sand	Dense	> 600
	Medium	200–600
	Loose	< 200
Sand	Compact	> 300
	Medium	100–300
	Loose	< 100
Cohesive soils		
Clay	Very stiff/stiff	150–600
	Firm	75–150
	Soft/silt	< 75
Peat and fill		N/A

(Figure 4.4). This is particularly the case in areas of clay, which are prone to considerable changes in volume as a result of changes in water content. This is linked strongly to the presence of trees. The extent to which mature trees extract water from the ground is considerable, and naturally this increases with the growth of the tree. This results in shrinkage of clay soils and is particularly noticeable in periods of dry weather, when the moisture is not replenished by rainfall. The consequence of this removal of moisture and the resulting shrinkage of the soil is that the foundations suffer from subsidence as the ground beneath them shrinks away. The effect is often one of rotation of the foundation rather than simple subsidence, as the area of soil beneath the building maintains a more stable moisture level. This is because the presence of the building prevents moisture loss from the ground by evaporation.

Problems occur not only because of the presence of trees, but also as a consequence of their removal. In areas where there have been established mature trees, the ground water level maintains a reasonable degree of equilibrium. The moisture level is regularised by the action of evaporation from the surface of the ground and moisture drawn by the trees. If the trees are then removed, the balance changes. In such circumstances the ground will retain more moisture than usual and in clay soils this will result in swelling of the ground. The swelling, in turn, causes uplift or 'heave' to buildings in close proximity. The effects of heave are similar to those of subsidence, but in reverse.

BS 5837: 2012 sets out recommendations for dealing with trees close to buildings. The effect of a tree's presence varies with species of tree and with soil type and characteristics.

PART 2

Figure 4.4
Proximity of trees.

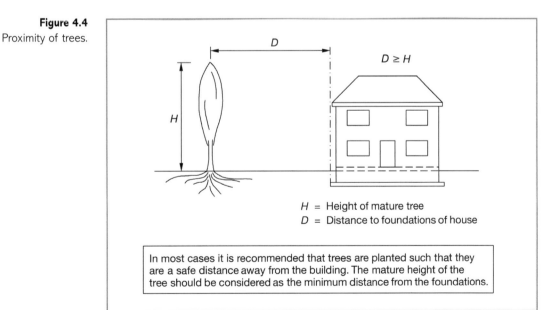

H = Height of mature tree
D = Distance to foundations of house

In most cases it is recommended that trees are planted such that they are a safe distance away from the building. The mature height of the tree should be considered as the minimum distance from the foundations.

REVIEW TASKS

■ A link has been noted between the soil type, its bearing capacity (pressure) and the loads from a building:

— How can this relationship be reflected by foundation design?
— What are the implications of foundation depth in terms of loadbearing capacity?
— What implication might the angle of repose of a soil type have in respect of foundation formation?

■ From the following list select the key features of cohesive and cohesionless soils:

— High proportion of voids between soil particles
— Compressibility
— High permeability
— Low permeability
— Variable volume with change in moisture content
— Ability to support itself during excavation.

■ Visit the companion website at www.palgrave.com/engineering/riley1 to view sample outline answers to the review tasks.

Contaminated ground

The risk of the presence of certain forms of contamination in the ground can be of great significance to the process of construction. Although it is not possible to consider this complex issue in depth within this book, it is useful to present the topic in overview. Contaminants that are likely to be found in the ground can pose problems in the form of potential risk to health or as aggressive substances to the materials used in construction. The Building Regulations define a contaminant as:

'any substance which is or could become toxic, corrosive, explosive, flammable or radioactive and likely to be a danger to health & safety'

Site investigation will be necessary to identify the nature and extent of contamination. This may have direct consequences for the viability of the proposed building project, and the requirements for remediation may be onerous. The Environmental Protection Act and Environment Act impose a responsibility for remedial action to negate the potential risks associated with land contamination. This can have very great consequences in terms of cost and time.

The main options available for treatment of the problem of contaminated ground are:

■ removal of the contaminated material to another location, normally a licensed site
■ provision of an impervious layer between the contamination and the ground surface, often termed 'cover technology'
■ biological or physical remediation to neutralise or remove the contaminant.

The types of contaminant typically found on sites that may be intended for building projects are set out in Table 4.3.

Table 4.3 Possible contamination risks on building sites.

Contaminant	Signs	Typical action
Metals	Affected vegetation Surface materials	Extraction of contaminated matter or cover technology
Organic compounds	Affected vegetation Surface materials	Biological remediation, extraction of affected matter or cover technology
Oil/tar	Surface materials	Extraction of affected matter
Asbestos/fibres	Surface materials	Extraction of affected matter or cover technology
Combustible materials	Surface materials Fumes/odours	Cover technology with venting facility or extraction of affected matter
Gases (methane/ carbon dioxide)	Fumes/odours	Barrier or cover technology
Refuse/waste	Surface materials Fumes/odours	Extraction of affected matter or cover technology

4.2 | Functions of foundations and selection criteria

Introduction

- After studying this section you should have developed an understanding of the nature and functions of foundations for low-rise construction.
- You should understand the implications of loading and ground conditions upon the selection of foundations.
- You should be able to identify and evaluate the factors that affect their choice and be able to select appropriate foundation solutions in a variety of scenarios.
- You should be aware of the merits and limitations of the various options.
- You should have an appreciation of the typical sequence of operations involved with their formation.

Overview

The foundations of a building form the interface between the structure and the ground which supports it. The primary function of the foundations is to transmit the building loads safely to the supporting strata, spreading them over a sufficient area to ensure that the safe loadbearing capacity of the soil is not exceeded. In addition, it must be ensured that the pressure on the ground at all points below the foundation is equal, in order to prevent differential movement or rotation. It is normal, therefore, to attempt to ensure that the centre of gravity of the loadings is located at the centre of the foundation area.

Functions of foundations

As previously noted, the primary functional requirement of a foundation is to transfer the loads from the building to the ground to ensure the stability of the structure (Figure 4.5). In order to accomplish this, a number of performance requirements

Figure 4.5
Load transfer through foundations.

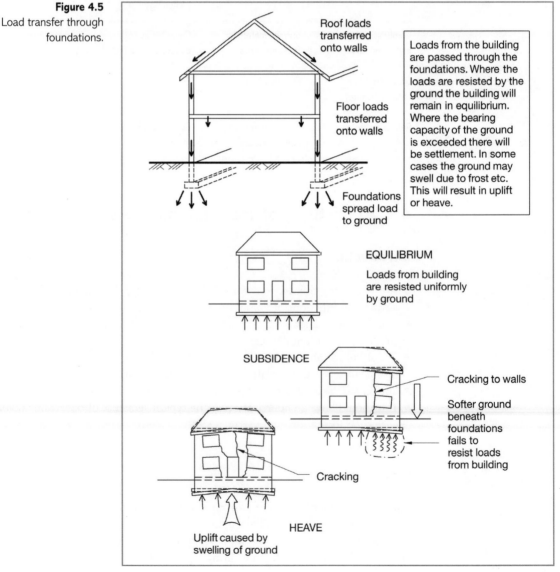

Roof loads transferred onto walls

Floor loads transferred onto walls

Foundations spread load to ground

Loads from the building are passed through the foundations. Where the loads are resisted by the ground the building will remain in equilibrium. Where the bearing capacity of the ground is exceeded there will be settlement. In some cases the ground may swell due to frost etc. This will result in uplift or heave.

EQUILIBRIUM

Loads from building are resisted uniformly by ground

SUBSIDENCE

Cracking to walls

Softer ground beneath foundations fails to resist loads from building

Cracking

HEAVE

Uplift caused by swelling of ground

must be achieved. These aspects of the performance of foundations are satisfied in different ways by differing foundation solutions. However, they are all essentially similar in the end result that is desired: to provide a stable durable base for the construction of the dwelling.

The following summarises the main aspects of foundation performance.

- The foundation must possess sufficient strength and rigidity to ensure that it is capable of withstanding the loadings imposed upon it without bending or suffering shear failure at the point of loading.
- It must be capable of withstanding the forces exerted by the ground, and must resist the tendency to move under such loads as those exerted by volume changes in the ground caused by frost and moisture action and so on. This is, in part, achieved by ensuring that the foundation is set at a sufficient depth below ground level, normally at least 1 metre.
- It must be inert, and not react with elements within the ground, such as soluble sulphates, which may degrade the material from which the foundation is formed.

Selection criteria

The basis for the selection of foundations is quite straightforward. It is important to remember that the performance of foundations is based on an interface between the loadings from the building and the supporting ground or strata. The nature and conditions of each of these may vary, and it is primarily as a result of these variations that some foundation solutions are more appropriate in certain circumstances than others. In all cases, however, the most economical solution will be selected, provided that it satisfies the performance requirements.

Factors related to ground conditions

In most cases the nature of the ground upon which dwellings are constructed is reasonably stable, level and of uniform composition. It is generally the case that the ground close to the surface is capable of supporting the relatively light loadings resulting from the construction of a dwelling. Thus, shallow forms of foundation are generally adopted. In some instances this is not the case, and the foundation solutions must be selected accordingly. The specific factors related to the nature of the ground that affect foundation selection are as follows:

Bearing capacity of the ground
In section 4.1 we considered the variability of soils and the degree to which their ability to carry loads differs. This is one of the key elements in the selection of appropriate foundations for all types of building.

Depth of good strata
Although the site upon which the dwelling is to be constructed may provide ground of appropriate bearing strength, this may be at a considerable depth below the

surface. In these circumstances the use of a shallow foundation form is unlikely to be efficient or cost-effective, and the adoption of a deeper foundation such as piles would be considered.

Composition of the ground

Ideally the construction of houses takes place on sites with uniform, stable ground conditions. As more and more sites have been utilised in recent years, the supply of these ideal sites has diminished. Hence it is becoming increasingly the case that sites for the construction of dwellings are less than ideal. The increasing trend to build on 'brownfield' sites, driven by the need to develop in a more sustainable way, has led to greater variability in ground composition. It is not uncommon to find houses being constructed on sites that feature areas made up of filled ground. Sites with lower than ideal bearing capacities may also be considered. These circumstances generally make the use of a traditional strip foundation unviable.

Ground level and gradients

It is rare for building sites to be truly flat and level. In most cases, however, the slight irregularities and changes in level of the ground do not pose significant problems. In cases where the variation in level is more extensive the design of the foundation will be affected. Several factors are significant in this scenario. Firstly, there is the need to provide a level base from which to build; this results in the need to step the foundation along the slope. Secondly, there is the potential for the entire structure to slide down the slope as a result of the action of gravity. In addition, there is the need to consider the implications of embedding parts of the building below ground level and/or creating deep sub-floor voids.

There are several potential design solutions to facilitate the construction of buildings on sloping ground (Figure 4.6). Each has merits and disadvantages. The first option is to provide a level foundation that cuts into the ground. This results in the need to excavate a pocket in the ground at the upper end of the building, while creating a deep sub-floor void at the lower end. A factor that must be taken into account here is the degree to which lateral loading will be applied to the walls of the building, where the ground level is high relative to the ground floor of the building. There is potential for the lateral load to cause deformation if it is of sufficient magnitude.

Lateral load is load that is applied from the side of a structural element. The implication of this load may be overturning, sliding or collapse of the element. When such loads are encountered because of ground levels being different on either side of a wall, this is often referred to as a 'retaining wall'. Its function is to retain the mass of earth on the one side of the wall.

One way of avoiding this is to adopt the second available option, which relies on the excavation of a pocket in the sloping ground and the formation of a retaining wall to protect the structural wall of the building. One of the implications of adopting either of these approaches is that the water that will run down the slope must be dealt with if penetration to the interior is not to be a problem.

The third approach is to create a stepped foundation that results in the floor level of the building changing as it passes up the slope. A benefit of this is that there is no need to excavate an excessive depth of soil. In addition, the level of **lateral load** applied to the walls is minimised. However, where such an approach is adopted, care must be taken to ensure that the foundation is of sufficient strength at the step to avoid fracture. This requires the incorporation of a sufficient overlap as detailed in Figure 4.7.

Figure 4.6
Building on slopes.

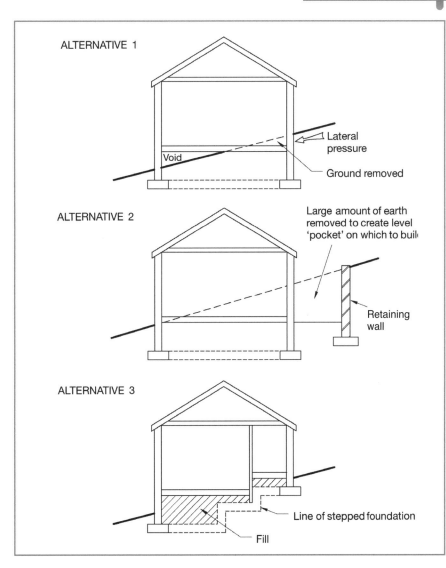

When the ground slopes it may make economic sense to step the foundation rather than have excessively deep trenches and excessively deep foundation brickwork. Figure 4.7 shows the dimensional needs at a step for compliance with the Building Regulations.

Factors relating to loads from the building

The form and type of the building to be erected dictate the nature of the loadings that are directed to the foundations, as well as their magnitude. In the case of low-rise housing the extent of the loading is relatively modest. This is one of the reasons for the widespread use of shallow foundations. As loads increase the need to provide deeper foundation solutions increases also. This is because the strata at a greater depth are more highly compacted, and generally, therefore, have the ability to with-

Figure 4.7
Steps in strip footing foundations.

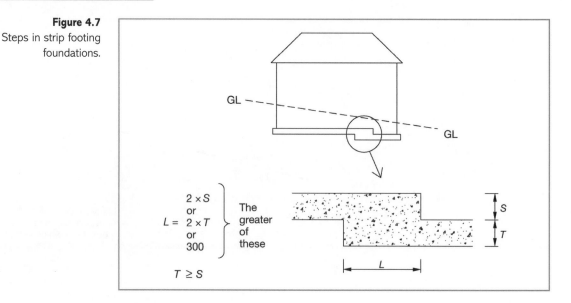

stand higher loads. By contrast, the strata close to the surface are generally less well compacted and less able to withstand loading.

4.3 | Types of foundation

Background and context

The nature of the foundation that is appropriate in a particular situation depends on a number of factors. The extent of the loading and the way in which it is applied to the foundation dictate the design, together with the nature of the supporting strata. Point loads on level sites are, for example, treated in a very different manner from uniformly distributed loads on sloping sites. Most forms of foundation are formed from concrete, either reinforced or unreinforced, although other forms such as steel grillage foundations are sometimes used in very specific circumstances.

In general, the foundation solutions available for the construction of dwellings are restricted to simple forms; these are normally defined within the classifications of shallow foundations and deep foundations.

REVIEW TASKS

■ With your present knowledge, and before you read the rest of this section, see how many different foundation forms you can mention.

■ Visit the companion website at www.palgrave.com/engineering/riley1 to view sample outline answers to the review tasks.

Shallow foundations forms

Introduction

■ After studying this section you should have developed an understanding of the nature and functions of shallow foundations for low-rise construction.
■ You should understand the nature and construction form of shallow foundations.
■ You should be able to identify and evaluate the factors that affect their choice.

Overview

The nature of the construction form of houses is such that the loads applied to the ground are relatively modest. In addition, they are generally distributed uniformly through the external loadbearing walls and any internal loadbearing walls. For this reason, the most popular choice of foundation type has historically been a shallow strip foundation. This is not always the case, however, and as the quality of available building land diminishes, so the possibility of foundation-related difficulties becomes more significant. A range of shallow foundation forms is available and the nature of the selected construction form, the likely loads and the bearing capacity of the ground will all be considered when making a selection.

Strip foundations

Mass concrete is the term used to refer to concrete that is cast into the trench without the use of reinforcement – literally a mass of concrete.

By far the most common form of foundation in use is the simple **mass concrete** strip foundation. Used where continuous lengths of wall are to be constructed, the strip foundation consists of a strip of mass concrete of sufficient width to spread the loads imposed on it over an area of the ground which ensures that overstressing does not take place. The distribution of the compressive loading within the concrete of the strip tends to fall within a zone defined by lines extending at 45 degrees from the base of the wall. Hence, to avoid shear failure along these lines, the depth of the foundation must be such that the effective width of the foundation base falls within this zone (Figure 4.8).

Strip foundations are intended for use where the loads from the building are relatively modest, as in the case of house building, and are distributed uniformly (Figure 4.9).

Since the external walls of dwellings transfer the structural loads to the foundations they are generally uniform in distribution. In some forms of construction the loads are transferred as localised or point loads (Figure 4.9). In such instances the loads are catered for by the adoption of isolated foundations at each point of load. This is rarely encountered in house building.

The traditional form of most house construction allows the use of shallow strip foundations in most instances (Figures 4.10 and 4.11). The width of the foundation depends on the safe bearing capacity of the soil and the level of loading applied. As discussed elsewhere, the width of the foundation is adjusted to provide a sufficient area of contact to prevent overstressing of the soil. The depth will be related to the depth of appropriate bearing strata, but will generally be around 1 metre so as to

Figure 4.8
Foundation design to
avoid shear failure.

The loads exerted on the central zone of the foundation are distributed within the concrete at a 45° angle. This results in the creation of a defined area of loading at the point of contact with the ground. The loads are resisted by the reaction force from the ground. If the width of foundation is such that this reaction is applied outside the zone of applied loading shear failure may result.

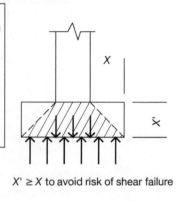

$X' \geq X$ to avoid risk of shear failure

In order to avoid the risk of shear failure the foundation is designed so that the foundation does not extend outside the zone of downward loading. The width of foundation is dictated by the strength of the soil and the extent of the applied load. (Area is adjusted to reduce pressure.) As the width increases, so the depth of the foundation material is increased so that there is no chance of shear failure.

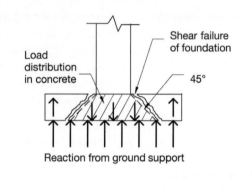

Load distribution in concrete

Shear failure of foundation

45°

Reaction from ground support

Figure 4.9
Uniform and point loads.

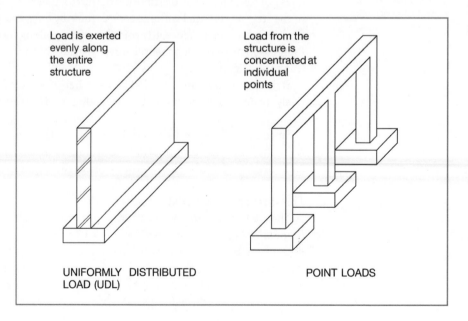

Load is exerted evenly along the entire structure

Load from the structure is concentrated at individual points

UNIFORMLY DISTRIBUTED LOAD (UDL)

POINT LOADS

Figure 4.10
Traditional strip footing
foundation.

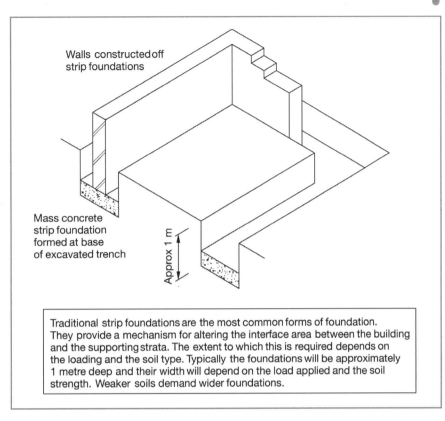

Walls constructed off
strip foundations

Mass concrete
strip foundation
formed at base
of excavated trench

Approx 1 m

Traditional strip foundations are the most common forms of foundation.
They provide a mechanism for altering the interface area between the building
and the supporting strata. The extent to which this is required depends on
the loading and the soil type. Typically the foundations will be approximately
1 metre deep and their width will depend on the load applied and the soil
strength. Weaker soils demand wider foundations.

PART 2

avoid the danger of frost-related heave and shrinkage of the soil close to the surface.
If the depth extends significantly below this level an alternative foundation form
may be necessary.

REVIEW TASKS
■ Explain how shear failure can occur in a strip footing foundation and how this
dictates certain dimensions of the concrete.

■ Using your existing knowledge of house construction identify which loads from
the building will need to be diverted to the foundations. By what route are the
loads transferred?

■ Visit the companion website at www.palgrave.com/engineering/riley1 to view
sample outline answers to the review tasks.

Excavation for shallow foundations

Excavation for the placement of foundations takes place after the building has been
'set out' to indicate the positions of trenches using profile boards and string lines.

The first stage of the process is the removal of 'oversite'. Following the removal
of the oversite, excavation of the foundation trenches will take place to the required

depth. On very small sites this may be undertaken by hand; however, the use of compact mechanical diggers makes mechanical excavation practical and economical on even modest sites. For larger sites it is the norm to use mechanised plant for excavation.

The excavation of foundation trenches follows an established sequence of operations, which ensures that they are formed to an appropriate standard and with required levels of safety for the building operatives.

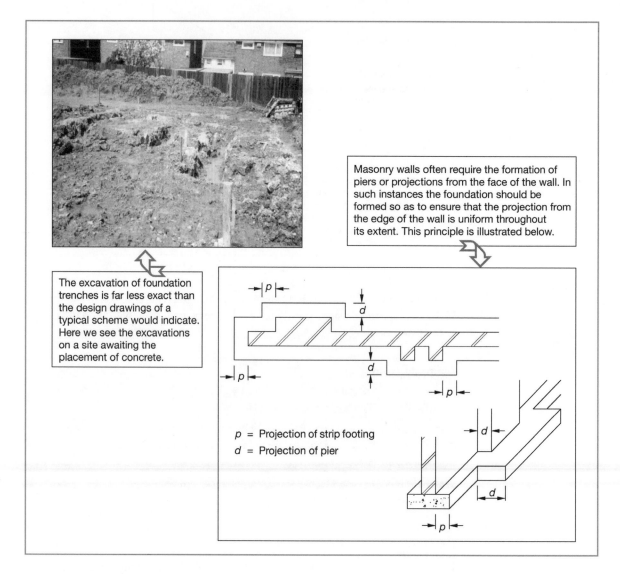

Masonry walls often require the formation of piers or projections from the face of the wall. In such instances the foundation should be formed so as to ensure that the projection from the edge of the wall is uniform throughout its extent. This principle is illustrated below.

The excavation of foundation trenches is far less exact than the design drawings of a typical scheme would indicate. Here we see the excavations on a site awaiting the placement of concrete.

p = Projection of strip footing
d = Projection of pier

Figure 4.11
Strip foundations.

REVIEW TASKS

■ The sequence of operations involved in excavation for the placement of foundations is important. From the following list of activities select the operations involved in the appropriate order:

- Excavation of trench
- Placement of concrete
- Levelling and ramming trench bottoms
- Setting out
- Excavation of oversite
- Timbering of trench sides
- Trimming of trench sides.

■ Visit the companion website at www.palgrave.com/engineering/riley1 to view sample outline answers to the review tasks.

As the loading exerted on the foundations increases or the loadbearing capacity of the ground decreases, so the required foundation width, to avoid overstressing of the ground, increases. In order to resist shear failure in wide foundations, they would be required to be very thick, for the reason previously described; this would normally be uneconomic. Hence, to reduce the required foundation thickness while still resisting failure in shear or bending, steel reinforcement is introduced to the regions subjected to tension and shear. Such a formation is termed a 'reinforced wide strip foundation' (Figure 4.12).

One of the practical problems in constructing walls off strip foundations at the depths required is the necessity to provide an excavation which is sufficiently wide to allow operatives to lay bricks in the region below ground level. Hence the trench is often excavated to a width which is in excess of that required, simply to provide a foundation of sufficient width. When working in cohesive soils, the sides of trenches are able to support themselves when excavated, without the use of earthwork support or foundation formwork. This means that a deep strip foundation (Figure 4.13) can be used.

This method of foundation formation is efficient and effective, functionally and in terms of time and cost on-site. The trench is excavated to the width necessary to satisfy loadbearing requirements and to a depth sufficient to avoid the zone affected by moisture variation and frost action. Concrete is then used to fill the excavation to near ground level, thus removing the need for many bricks or blocks to be laid below ground level.

REVIEW TASKS

■ Why is a deep strip footing only suited to certain types of soil?

■ What is the implication of low soil strength on strip foundation design?

■ Visit the companion website at www.palgrave.com/engineering/riley1 to view sample outline answers to the review tasks.

Figure 4.12
Design of wide strip foundations to avoid failure in shear and bending.

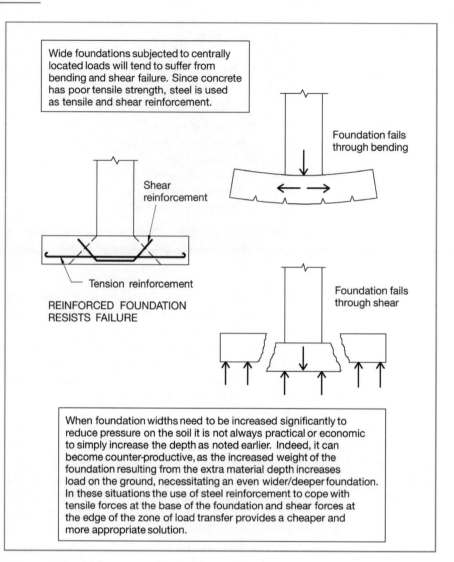

Wide foundations subjected to centrally located loads will tend to suffer from bending and shear failure. Since concrete has poor tensile strength, steel is used as tensile and shear reinforcement.

Foundation fails through bending

Shear reinforcement

Tension reinforcement

REINFORCED FOUNDATION RESISTS FAILURE

Foundation fails through shear

When foundation widths need to be increased significantly to reduce pressure on the soil it is not always practical or economic to simply increase the depth as noted earlier. Indeed, it can become counter-productive, as the increased weight of the foundation resulting from the extra material depth increases load on the ground, necessitating an even wider/deeper foundation. In these situations the use of steel reinforcement to cope with tensile forces at the base of the foundation and shear forces at the edge of the zone of load transfer provides a cheaper and more appropriate solution.

Pad foundations

Pad foundations are rarely used in the construction of houses. They are discussed here only in overview to provide a general awareness of their existence and form, as they may be used from time to time to support elements such as props to car ports, porticos and so on.

In some instances, such as when constructing framed buildings, the loads from the building are exerted upon the foundations in the form of concentrated point loads. In such cases, independent pad foundations are normally used. The principle of pad foundation design is the same as that of strip foundations, that is, the pad area will be of sufficient size to safely transfer loadings from the column to the ground without exceeding its loadbearing capacity. The pad will generally be square, with the column loading exerted centrally to avoid rotation, and is normally formed from reinforced concrete; although other formations such as steel are possible, they

Pad foundations are used extensively in the construction of industrial and commercial buildings that employ structural frames to transmit loads rather than the loadbearing fabric associated with many houses. The use of a structural frame results in point loads on the foundations.

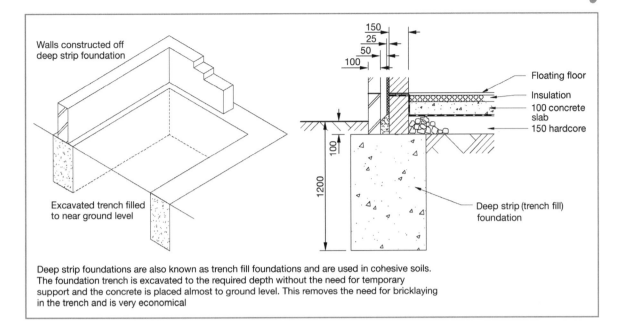

Walls constructed off deep strip foundation

Floating floor

Insulation

100 concrete slab

150 hardcore

Excavated trench filled to near ground level

Deep strip (trench fill) foundation

Deep strip foundations are also known as trench fill foundations and are used in cohesive soils. The foundation trench is excavated to the required depth without the need for temporary support and the concrete is placed almost to ground level. This removes the need for bricklaying in the trench and is very economical

Figure 4.13
Deep strip foundations.

PART 2

are rare in modern construction. As a result of the concentration of loading applied to pad foundations, it is vital that they are reinforced sufficiently against shear as well as bending.

The nature of the location of the loadbearing columns at the pad foundation can vary greatly in different situations. Figures 4.14 and 4.15 illustrate some of the more common methods.

Figure 4.14
Pad foundations.

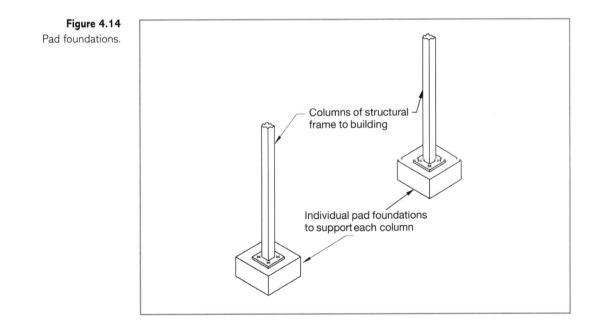

Columns of structural frame to building

Individual pad foundations to support each column

Figure 4.15
Standard column fixing at
pad foundations.

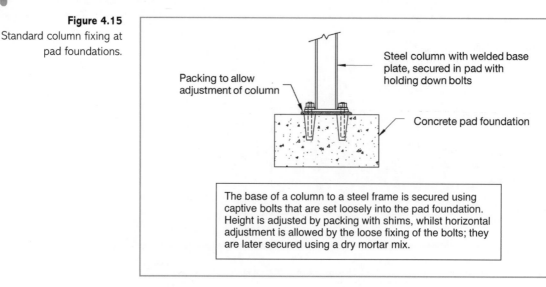

Packing to allow
adjustment of column

Steel column with welded base
plate, secured in pad with
holding down bolts

Concrete pad foundation

The base of a column to a steel frame is secured using
captive bolts that are set loosely into the pad foundation.
Height is adjusted by packing with shims, whilst horizontal
adjustment is allowed by the loose fixing of the bolts; they
are later secured using a dry mortar mix.

Raft foundations

Raft foundations (Figure 4.16) are constructed in the form of a continuous slab,
extending beneath the whole building, and formed in reinforced concrete. Hence
the loadings from the building are spread over a large area and avoid overstressing
the soil. The foundation acts, quite literally, like a raft. Inevitably, however, there
tends to be a concentration of loading at the perimeter of the slab, where external
walls are located, and at intermediate points across the main surface, resulting from
loadings from internal walls and so on. Hence the raft is thickened at these locations
and/or is provided with extra reinforcement. At the perimeter (Figure 4.17), this
thickening is termed the 'toe' and fulfils an important secondary function in
preventing erosion of the supporting soil at the raft edges. If not prevented under-
mining of the raft could occur, with serious consequences.

Foundations of this type are generally utilised in the construction of relatively
lightweight buildings, such as houses, on land with low bearing capacity. One of the
advantages of this form of foundation is that if settlement occurs, the building
moves as a whole unit, differential movement is prevented, hence the building
retains its integrity. Raft foundations are used commonly in the construction of
timber framed houses.

REVIEW TASKS

■ There are clearly several design solutions for the provision of shallow foundations to dwellings. You should now have developed an appreciation for the
criteria upon which they are selected. Using these criteria, select a foundation
type from the following list of options for each of the situations listed below.

Options: Surface raft; Strip; Wide strip; Deep strip; Pad

Situation 1   The construction of a two-storey dwelling in an area of solid rock
substrata

Situation 2 The construction of a two-storey dwelling in an area of firm stiff clay

Situation 3 The construction of a two-storey dwelling in an area of soft clay

Situation 4 The construction of a two-storey dwelling in an area of filled ground

■ Visit the companion website at www.palgrave.com/engineering/riley1 to view sample outline answers to the review tasks.

Figure 4.16
Raft foundations.

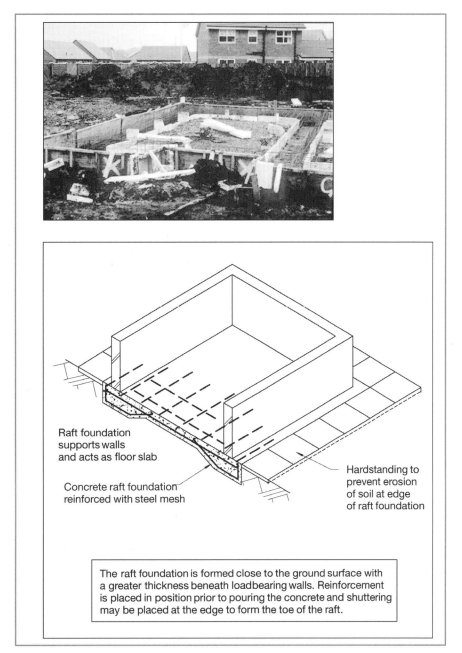

Raft foundation supports walls and acts as floor slab

Concrete raft foundation reinforced with steel mesh

Hardstanding to prevent erosion of soil at edge of raft foundation

The raft foundation is formed close to the ground surface with a greater thickness beneath loadbearing walls. Reinforcement is placed in position prior to pouring the concrete and shuttering may be placed at the edge to form the toe of the raft.

PART 2

Figure 4.17
Raft edge details.

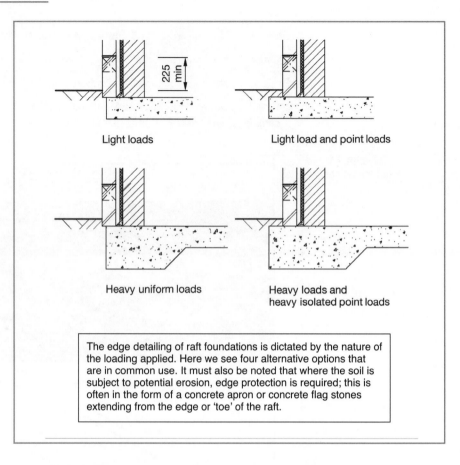

Light loads

225 min

Light load and point loads

Heavy uniform loads

Heavy loads and
heavy isolated point loads

The edge detailing of raft foundations is dictated by the nature of
the loading applied. Here we see four alternative options that
are in common use. It must also be noted that where the soil is
subject to potential erosion, edge protection is required; this is
often in the form of a concrete apron or concrete flag stones
extending from the edge or 'toe' of the raft.

Deep foundation forms

Introduction

- After studying this section you should have developed an understanding of the nature and functions of deep foundations to low-rise construction.
- You should understand the nature and construction form of deep foundations.
- You should be able to identify and evaluate the factors that affect their choice.

Overview

The use of deep foundation forms for the construction of dwellings is relatively uncommon. The expense of deep foundation formation is one reason for this. However, the main reasons are functional, since the relatively low loads exerted by dwellings are normally dealt with effectively by shallow forms, such as strip foundations. The increasing need to develop sites with less favourable ground conditions has resulted in an increased use of deeper foundation forms. These are needed to cope with sites where ground quality close to the surface is poor or variable. Hence in some circumstances deeper foundation forms are essential.

Piles

Piles are often described as columns within the ground, since the basis upon which they work is similar to that of a traditional column, in that they transfer loadings from a higher level to a loadbearing medium at a lower level. They can be categorised in two ways (Figure 4.18): by the way in which they are installed, or by the way in which they transfer their loads to the ground (Figure 4.19). Hence the following classifications are used to define pile types:

- Definition by installation method
  - Displacement piles (driven)
  - Replacement piles (bored)
- Definition by load transfer mechanism
  - Friction piles
  - End-bearing piles

Figure 4.18
Categories of pile.

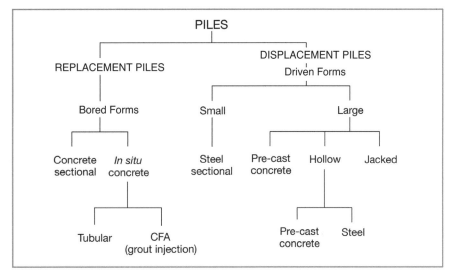

Displacement piles

Displacement piles are set into the ground by forcing or driving a solid pile or a hollow casing to the required level below ground, thus displacing the surrounding earth. In the case of solid piles, **precast** piles of required length may be driven into the ground using a driving rig, or alternatively the rig may be used to drive a series of short precast sections, which are connected as the work proceeds. The use of the second of these methods is by far the most efficient, since the length which is required may vary from pile to pile. Thus the use of one-piece precast piles inevitably necessitates adjustment of length on-site. Trimming or extending of one-piece piles is a difficult task on-site.

The difficulties of providing piles of exactly the right length, together with the danger of damage to the pile resulting from the percussive driving force have resulted in the adoption of the use of driven shell, or casing, piles. With this method

Reinforced concrete building elements that are used in the construction of houses may be made on-site by casting concrete *in situ*. Alternatively they may be made in a factory environment and installed as finished components; items such as lintels and so on may take this form. Such elements are referred to as **precast**.

PART 2

Figure 4.19
Load spread from shallow
and pile foundations.

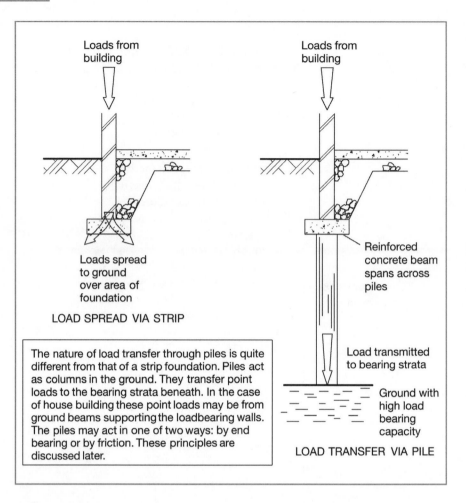

Figure 4.19
Load spread from shallow and pile foundations.

(Figure 4.20), a hollow shell or casing is driven into the ground, using a percussive rig, in a number of short sections; concrete is then poured into the void as the casing is withdrawn, the steel reinforcement having already been lowered into the hole. By this method, an *in-situ* pile is formed in the ground. The vibration of the pile casing, as it is withdrawn from the ground, results in the creation of a ridged surface to the pile sides, thus taking the most advantage of any frictional support provided by the ground.

Whether in the form of a solid pile or a hollow casing, the driving of the pile/casing is aided by the use of a driving toe or shoe, often in the form of a pointed cast-iron fitting at the base of the pile to allow easier penetration of the ground. These methods of installation have several disadvantages, in that considerable levels of noise and vibration are generated as a result of the driving operation. Hence they are generally considered unsuitable for congested sites, where adjacent buildings may be structurally affected, or areas where noise nuisance is undesirable. However, they may be used to good effect in consolidation of poor ground, by compressing the earth around the piles.

Figure 4.20
Installation of
displacement piles.

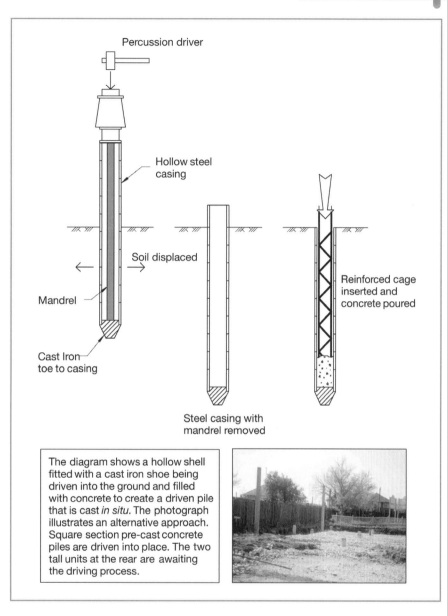

The diagram shows a hollow shell fitted with a cast iron shoe being driven into the ground and filled with concrete to create a driven pile that is cast *in situ*. The photograph illustrates an alternative approach. Square section pre-cast concrete piles are driven into place. The two tall units at the rear are awaiting the driving process.

Replacement piles

Unlike displacement piles, replacement piles are installed by removing a volume of soil and replacing it with a load-supporting pile (Figure 4.21). The holes are bored either by using a hollow weighted grab, which is repeatedly dropped and raised, removing soil as it does so, or by using a rotary borer or auger. As the excavation progresses, the sides are prevented from collapsing by introducing a shell, or casing, normally made from steel, or by the use of a viscous liquid called bentonite. The bentonite is then displaced by concrete as it is poured into the excavation and is stored for further use. This method is quieter than the displacement method and does not result in damage to surrounding buildings.

Figure 4.21
Installation of
replacement piles.

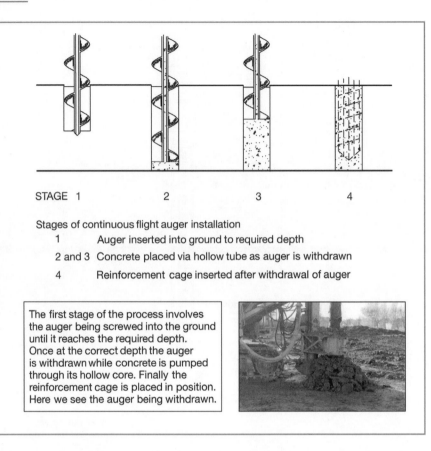

STAGE 1   2   3   4

Stages of continuous flight auger installation
1   Auger inserted into ground to required depth
2 and 3 Concrete placed via hollow tube as auger is withdrawn
4   Reinforcement cage inserted after withdrawal of auger

The first stage of the process involves
the auger being screwed into the ground
until it reaches the required depth.
Once at the correct depth the auger
is withdrawn while concrete is pumped
through its hollow core. Finally the
reinforcement cage is placed in position.
Here we see the auger being withdrawn.

The methods of installation of replacement piles described above, however, have largely been superseded by the introduction of **continuous flight auger**, or grout injection piling (Figure 4.22). This method is highly efficient, with exceptionally fast installation possible. A continuous auger of the required length is mounted on a mobile rig and is used to excavate the pile void. The central shaft of the auger is hollow, and is connected to a concrete pumping system, which allows the concrete to be placed as the auger is withdrawn. Hence there is no need to provide temporary support to the sides of the excavation. Once the auger is withdrawn, a steel reinforcement cage is forced into the concrete from above. Thus, the pile is formed quickly and efficiently with little disturbance to surrounding areas.

Continuous flight auger (CFA) piling uses a long auger (drill) to form piles in the ground quickly and quietly.

Friction piles and end-bearing piles

The method by which the pile transfers its load to the ground depends upon the nature of the design of the pile and is dictated to a large extent by the nature of the ground in which it is located. Piles are used as a form of foundation in a variety of situations, each imposing differing demands upon pile design and creating restrictions on the ways in which the piles act. Although there are a great variety of situations which may necessitate the use of piles, the following are some of the most common:

Figure 4.22

Installation of continuous flight auger (CFA) piles.

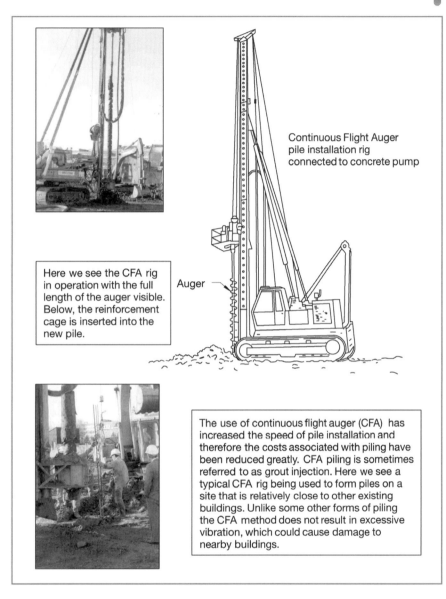

Continuous Flight Auger pile installation rig connected to concrete pump

Auger

Here we see the CFA rig in operation with the full length of the auger visible. Below, the reinforcement cage is inserted into the new pile.

The use of continuous flight auger (CFA) has increased the speed of pile installation and therefore the costs associated with piling have been reduced greatly. CFA piling is sometimes referred to as grout injection. Here we see a typical CFA rig being used to form piles on a site that is relatively close to other existing buildings. Unlike some other forms of piling the CFA method does not result in excessive vibration, which could cause damage to nearby buildings.

- Where insufficient loadbearing capacity is offered by the soil at a shallow depth, but sufficient is available at a greater depth
- Where the nature of the soil at a shallow depth is variable and performance is unpredictable, such as in areas of filled land
- Where soils at shallow depths are subject to shrinkage or swelling due to seasonal changes
- Where buildings or elements are subjected to an uplifting force, and require to be anchored to the ground.

It can be seen that in some of these instances one form of pile may be obviously more appropriate than another as a result of the way in which they act. End-bearing

Figure 4.23
End-bearing and
friction piles.

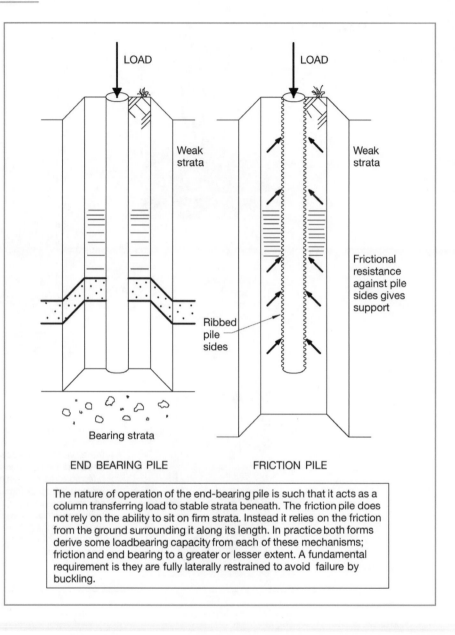

LOAD

LOAD

Weak
strata

Weak
strata

Frictional
resistance
against pile
sides gives
support

Ribbed
pile
sides

Bearing strata

END BEARING PILE

FRICTION PILE

The nature of operation of the end-bearing pile is such that it acts as a
column transferring load to stable strata beneath. The friction pile does
not rely on the ability to sit on firm strata. Instead it relies on the friction
from the ground surrounding it along its length. In practice both forms
derive some loadbearing capacity from each of these mechanisms;
friction and end bearing to a greater or lesser extent. A fundamental
requirement is they are fully laterally restrained to avoid failure by
buckling.

piles act by passing through unsuitable strata to bear directly upon soil with adequate
bearing capacity, while friction piles are supported by the effects of friction from the
ground to the sides of the pile throughout its length (Figure 4.23); hence the formation of ridges to the sides of piles as described earlier. In practice, all piles derive
their support from a combination of these factors.

Connections to piles

The processes of installation of piles result in the tops of the piles being far from
perfectly level and true; hence a loading platform must be created to take the loads

Figure 4.24
Pile cap configurations.

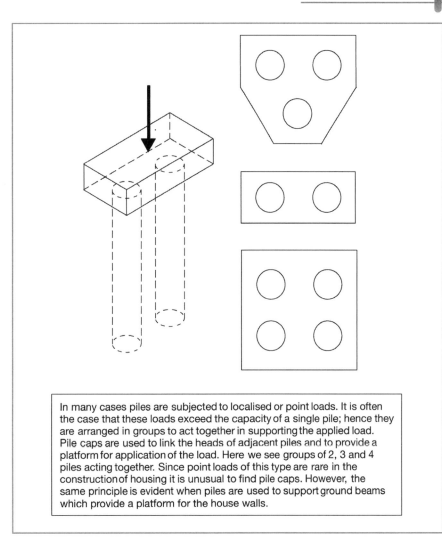

In many cases piles are subjected to localised or point loads. It is often the case that these loads exceed the capacity of a single pile; hence they are arranged in groups to act together in supporting the applied load. Pile caps are used to link the heads of adjacent piles and to provide a platform for application of the load. Here we see groups of 2, 3 and 4 piles acting together. Since point loads of this type are rare in the construction of housing it is unusual to find pile caps. However, the same principle is evident when piles are used to support ground beams which provide a platform for the house walls.

PART 2

A **ground beam** is a reinforced concrete beam cast at or just below ground level. It will generally be supported by piles at each end and its function is to provide support for the uniformly distributed loads associated with external walls or cladding.

from the building and transmit them safely to the piles. This platform is known as a pile cap (Figure 4.24), which is formed in reinforced concrete and may transfer loads to a single pile or a group of piles. The caps are normally loaded by the columns of the building and may also take loads from the walls via a **ground beam**, as shown in Figure 4.25.

Where houses are to be built on ground that is of poor loadbearing capacity or where there are particular problems with shrinkable clay, piles may be considered as the best alternative. However, piles are generally associated with the application of point loads such as from the columns of a framed building. In order to allow the transfer of uniformly distributed loads (UDLs) from the walls of a traditionally built dwelling, we must provide a suitable interface between the wall and the piles. This is effected by using a reinforced concrete ground beam. The beam spans the piles to form a continuous support for the construction of the walls. One potential problem

Figure 4.25
Short bored pile
foundations.

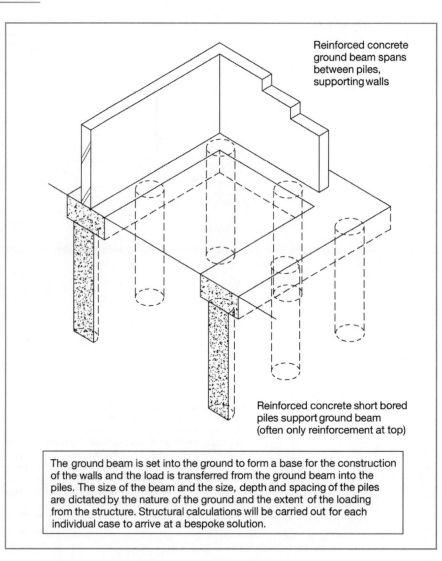

Reinforced concrete
ground beam spans
between piles,
supporting walls

Reinforced concrete short bored
piles support ground beam
(often only reinforcement at top)

The ground beam is set into the ground to form a base for the construction of the walls and the load is transferred from the ground beam into the piles. The size of the beam and the size, depth and spacing of the piles are dictated by the nature of the ground and the extent of the loading from the structure. Structural calculations will be carried out for each individual case to arrive at a bespoke solution.

in adopting this approach in shrinkable clay soils is that the volume of the ground may increase with moisture content as well as shrinking, depending on the time of year. This would cause heave or uplift on the underside of the ground beam, as this is in contact with the ground at a shallow level. The problem is avoided by the utilisation of a compressible board laid between the ground and the underside of the ground beam.

In the construction of dwellings it is unusual for there to be a need for piles to extend to great depths, since the loads are generally quite low. Hence one of the most common forms of piling used is the short bored pile (Figures 4.25 and 4.26).

One of the recent innovations in housing building has been the development of precast or prefabricated foundation systems. These rely on the use of piles, or piers and reinforced concrete ground beams. The piles are placed *in situ* to reflect the layout of the house walls. The precast ground beams are then placed to provide a

Figure 4.26
Reinforced concrete
ground beam.

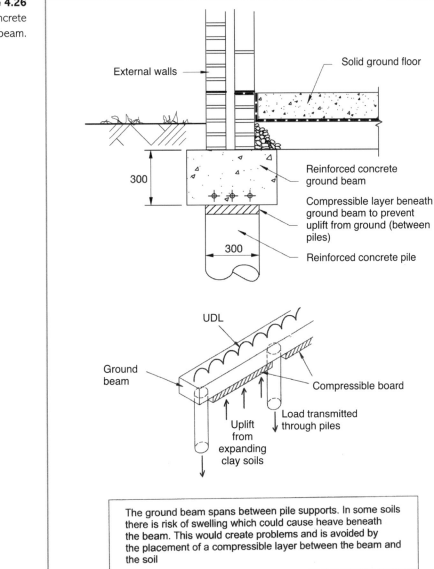

The ground beam spans between pile supports. In some soils there is risk of swelling which could cause heave beneath the beam. This would create problems and is avoided by the placement of a compressible layer between the beam and the soil

sound base for the construction of the walls. The structural principle of this system is the same as a traditional pile and beam system. However, the use of precast beams accelerates the construction process by removal of the activities associated with formwork and concrete placement, and removal of the time requirement for the concrete to cure. One of the benefits of this system over traditional mass concrete foundations is that it results in far less wasted, excavated material. As such, less material is directed to landfill and therefore the technique is considered more sustainable than traditional forms. As a result of the cost of landfill tax, there are also significant financial benefits.

Typical details for such a system are illustrated in Figures 4.27 and 4.28.

Figure 4.27
Precast foundation system.

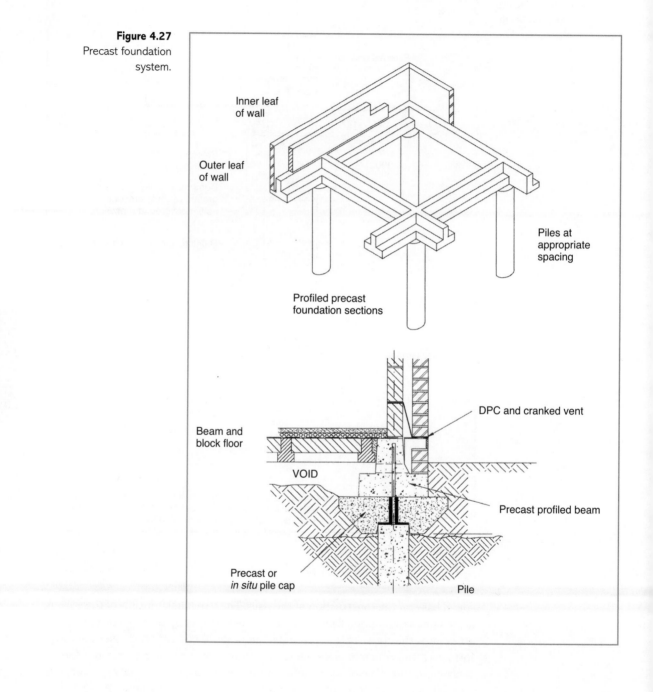

Figure 4.28
Precast foundations in use.

REVIEW TASKS

■ When dealing with domestic-scale construction, deep foundation forms tend to be limited to piles. With reference to this form of foundation, consider the following questions. Remember: there may be more than one correct answer!

1. In which of the following situations may piles be used as a foundation solution?

   (a) Where there is firm non-cohesive soil at depths close to the surface of the ground
   (b) Where the ground is made of firm clay
   (c) In areas of filled ground
   (d) In areas of peaty ground
   (e) In areas of mining subsidence.

2. Grout injection or continuous flight auger (CFA) piles are commonly used. Select the features which you consider to be the advantages of this form of piling:

   – Speed
   – Economy
   – Low disturbance
   – Ground consolidation
   – Loadbearing capacity.

3. Driven piles also have advantages. Select the features which you consider to be the advantages of this form of piling:

   – Speed
   – Economy
   – Low disturbance
   – Ground consolidation
   – Loadbearing capacity.

■ Visit the companion website at www.palgrave.com/engineering/riley1 to view sample outline answers to the review tasks.

COMPARATIVE STUDY: FOUNDATIONS

Option	Advantages	Disadvantages	When to use
Strip	– Cheap – Familiar technology – No need for specialised plant	– Requires formwork in non-cohesive soils – Access for bricklaying in trench	– Continuous walls on ground with reasonable bearing capacity – Used in most cases for construction of dwellings
Deep strip	– No need for formwork – Reduced bricklaying below ground – Fast – Cheap – No need for specialised plant	– Unsuitable for cohesionless soils – May be susceptible to movement in shrinkable clay	– In cohesive soils where depth of foundation is not excessive and loads are uniform and continuous
Wide strip	– Allows increased bearing on low-strength soils – Avoids excessive depth of foundation as width increases	– Requires reinforcement to resist bending and shear – Can become uneconomic compared with raft foundation as width increases	– Continuous walls on ground with low bearing capacity
Piles	– Economic when bearing stratum is deep – Avoids shrinkable clay zone – Allows use of land unsuitable for traditional foundations	– Need for special plant – Potential for vibration or damage from driven piles – More complex structural design required	– Used with ground beams to support house walls where loadbearing stratum is deep below ground level – Where shrinkable clay layer close to the surface must be avoided – Used singly or as clusters to support isolated point loads
Raft	– Economic as it combines foundation and floor slab – Shallow excavations avoid need for costly plant	– Requires reinforcement to be designed – Can suffer from edge erosion if not protected	– Used for relatively light loads on ground with poor or variable loadbearing capacity – Popular with timber frame construction – Often used on filled or brownfield sites
Precast system	– Dry process increases speed – Fast – Suitable for poor bearing conditions – Sustainable solution	– Unfamiliar – Costly compared to 'traditional' forms (although cost is mitigated by landfill saving) – Lead-in time	– Used with ground beams to support house walls where loadbearing stratum is deep below ground level – Where shrinkable clay layer close to the surface must be avoided – Used singly or as clusters to support isolated point loads

PART 2

5

Walls below ground

AIMS

After studying this chapter you should be able to:

- Understand the nature of the environment in which walls below ground are required to function
- Appreciate the factors that affect the performance of walls below ground level and the materials used for their construction
- Identify and apply criteria for the selection of appropriate design solutions, and be familiar with the form and construction details of the various options

This chapter contains the following sections:

5.1   Functional requirements of walls below ground
5.2   Options for walls below ground
5.3   Entry of services

INFO POINT

- Building Regulations Approved Document A, Structure (2004 including 2010 amendments)
- BS 5997: Guide to British Standard codes of practice for building services (1980)
- BS 6515: Specification for polyethylene damp-proof courses for masonry (1984)
- BS 8103: Structural design of low-rise buildings. Code of practice for masonry walls in housing (2005)
- BS 8110: Structural use of concrete (1997)
- BS 8215: Code of practice for design and installation of damp-proof courses in masonry construction (1991)

5.1 | Functional requirements of walls below ground

Introduction

■ After studying this section you should be aware of the main functions of walls below ground.
■ You should be able to identify the key issues affecting the selection of appropriate design solutions.
■ You should have developed an appreciation of the variable factors linked to the selection of sites for building that may have an effect on the performance of walls below ground.

Overview

Hydrostatic pressure is the term used to refer to the pressure upon moisture in the ground associated with its depth. As the depth increases, water is driven under pressure to areas such as voids and cavities that are at lower pressure.

The sections of walls below ground level provide an important link between the superstructure of the building and the foundations. The loads from the building must be transferred safely through these walls and the passage of moisture from the ground must be checked to prevent entry to the building. At the same time, the walls are subjected to lateral loadings from the ground surrounding them and, depending on the relative levels of the ground on each side, may be subject to **hydrostatic pressure** driving moisture through the structure. Within this section we shall consider the structural functions of the walls below ground and the implications of selection of various options upon the construction process and the cost of building.

The main functional requirements of these sub-ground walls are structural stability, the exclusion of moisture and durability (Figure 5.1). Since the walls are below ground they are rarely seen; hence appearance is not significant.

Durability of walls below ground

As previously noted, the environment in which some elements of the structure are sited below ground can be hostile to the materials of construction. This imposes some restrictions on the nature of construction of walls below ground level. These walls are subjected to ground moisture for much of the time. The extent of this depends upon the specific nature of the soil and the height of the water table, but in most cases the walls below ground are in a wet environment. Hence materials with low porosity are desirable; these are less likely to suffer deterioration as a result of the freezing of the water close to the surface of the ground in periods of cold weather. Porous materials tend to absorb ground moisture, which expands on freezing causing spalling and friability of the material.

The walls in these areas are also subjected to high pressures, both from the building above ground and the ground itself. The lateral force exerted by the mass

PART 2

Figure 5.1
Functions of walls below ground

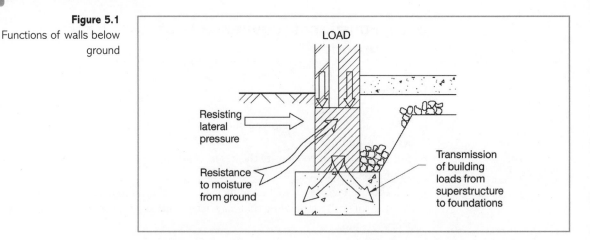

of earth which surrounds the walls can have a considerable compressive effect, particularly in the case of cavity walls (Figure 5.2).

In order to resist the effects of this loading it is common practice to fill the cavity below ground level with weak mix concrete, or to construct the walls up to ground level from solid brick or block, without the cavity. The blocks that are used in this area are very durable, being of high density with low porosity. They are very different from the form of blockwork which is generally used above ground. This type of block is known as 'trenchblock' and is specifically manufactured for use below ground level.

Another factor that must also be considered is the potentially deleterious nature of the soil in which the walls are to be placed. In clay soils there is the potential for the action of soluble sulphates to deteriorate concrete. Thus a full site investigation to identify materials that have the potential to adversely affect the durability of the walls is essential. In the case of sulphate attack in concrete and the cement mortar used to bond the brick or block wall, the effect is to cause expansion, friability and loss of strength. The effects are irreversible and the implications for durability are considerable.

Figure 5.2
The effects of soil pressure on walls below ground.

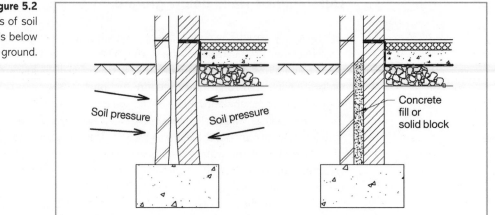

Building Form and Function

Plate 1: Traditional dwellings in a tropical climate
The form of these dwellings is driven by the locally available materials and the tropical climate. They are 'thermally light' buildings that can react quickly to temperature changes and the large openings allow cooling breezes to have maximum effect.

Plate 2: Thermally heavy building
This building uses locally available material to create a thermally heavy structure, capable of storing heat within the fabric and providing insulation from external temperature changes.

Traditional Construction

Plate 3: Precast proprietary concrete foundations
These can be used in association with many methods of house construction. Here they are seen combined with traditional masonry construction.

Plate 4: Beam and block ground floor
This image illustrates the detail at the junction of the external wall and ground floor. Beam and block floors are commonly used where there is a deep floor void and they provide a cost-effective, robust construction solution. Note the soil pipe passing through the floor.

Plate 5: Service pipes
The soil pipe passes through the ground floor and exits through the external wall below ground level. These must be placed in position at an early stage of construction.

Plate 6: Pre-formed cavity closer
The requirement to provide an insulated and fire-resistant closure at openings in cavity walls is generally dealt with by the installation of a proprietary closure profile. As can be seen here, they also provide a former for the construction of openings in the external wall.

Plate 7: Traditional cavity wall
Here we see the detail of the cavity wall adjacent to a window opening. Note the position of the insulation, which leaves a partial cavity. The stainless steel cavity tie is also visible, which will link the inner and outer leaves of the external wall.

Plate 8: Lintel detail
The window opening is formed with a galvanised steel lintel, which supports the outer and inner leaves of brick and block. The insulated cavity closer separates the two leaves at the reveal and provides a fire-resistant closure.

Plate 21: Roof structure
The structure of the roof relies on the actions of a series of ties and struts that make up the trussed rafter sections. These are connected using galvanised steel 'gang plates'.

Plate 22: Timber frame nearing completion
The wall and roof structures are seen here nearing completion. Note the 'gable ladder' at the edge of the roof that allows a projection to receive the verge boards.

Steel Framed Construction

Plate 23: Steel frame
The structural frame of the house is formed using lightweight steel, which is then insulated and clad to provide a weather-resistant enclosure.

Timber Framed Construction

Plate 16: Mass concrete strip foundation

The use of poured concrete strip foundations is still the most common form in domestic construction.

Plate 17: Timber frame at DPC level

The timber frame that will cater for the structural loads of the house is placed *in situ* and connected to the foundations before any work commences on the outer leaf or cladding. Here we see the frame, covered with breather paper and reflective sheet, along with the lower section of the external wall with the DPC clearly visible.

Plate 18: Upper floor construction

These 'open-web' joists are used to reduce weight and minimise use of materials. The upper and lower flange sections are connected using galvanised steel struts. Such joists also have the added benefit of easy accommodation of services.

Plate 19: Trussed rafter roof

The trussed rafter roof is assembled and provided with temporary and permanent bracing to ensure stability. The gable peak is placed in position in sections that match the lower part of the timber frame.

Plate 20: Party wall junction

Two separate frame sections are positioned adjacent to each other to create the party wall detail. They are connected using steel straps, but maintain a cavity to prevent noise transmission.

Plate 14: Roof truss fixings
The roof trusses must have a sound connection to the external walls to ensure effective transfer of loads. This is effected by connecting individual trusses to a continuous timber wall-plate using galvanised steel saddle units.

Plate 13: Roof formation
The roof is formed using trussed rafter sections that are manufactured off-site in a range of standard profiles. This increases the speed of construction and reduces waste on-site considerably.

Plate 15: Nearing completion
The property is nearing completion. Traditional design features have been included to increase their attractiveness to potential purchasers. The external landscaping is still to be completed and the exposed foundations and below-ground drainage are clearly visible.

Plate 9: Traditional timber upper floor
The traditional form of upper floor for dwellings adopts the use of timber joists seated in pockets in the external wall or, as illustrated here, using hangers. The use of hangers improves air-tightness in the external envelope. Note the trimmed opening that will accommodate the stairs.

Plate 10: Upper floor construction
This example of upper floor construction features timber I-section joists. These allow the sustainable production of lightweight sections that can deal with relatively long spans. Note the use of double joists and galvanised steel hangers to connect the joists around the trimmed stair opening.

Plate 11: Level threshold
The creation of a level threshold at the front door requires careful detailing to ensure that water does not pass into the building. The drainage channel is protected during construction by using a simple plywood covering sheet.

Plate 12: External wall construction
The external walls are constructed in a series of 'lifts'. As the height of the wall increases the scaffolding is extended to provide a safe working platform. The scaffold is often tied into the external walls to ensure stability. Note the 'toe boards' on the scaffold that prevent items from falling or being kicked off the scaffold.

Plate 24: Insulation applied to the frame
The steel frame is completed for this dwelling, which is on a sloping site. The frame is clad with insulation board to create a 'warm frame' enclosure.

Plate 25: Frame nearing completion
The completed frame is seen here with the cladding applied to the walls. Note the protruding lattice-truss roof sections.

Plate 26: Connection details
The steel frame sections are braced for stability using diagonal straps. The steel frame units are formed using C-section studs formed from lightweight galvanised steel.

Plate 27: Upper floor structure
The upper floors are formed using lattice joists that span between the steel frame units. They offer excellent strength-to-weight ratio.

Plate 28: Upper floor detail
The floor is seen here from the underside. The lattice joists are clearly visible along with their connections to the supporting walls. Note the detail of the internal door opening to the left of the picture.

Plate 29: Roof structure
The pitched roof structure is visible in this picture with a lattice girder forming the main ridge section. This property is designed to feature rooms within the roof space. Hence, the absence of trussed rafter units.

Plate 30: External cladding
The completed frame is clad in this case with reclaimed brick to give a traditional external appearance.

Alternative Approaches

Plate 31: Hempcrete
This dwelling is constructed from hempcrete, a mixture of hemp hurds (woody core) and a hydroscopic (breathing) binder, usually lime based which produces an insulating and heat-storing material that is used to build the walls.

Plate 32: Insulated concrete formwork
In this system, walls are formed by placing high-strength concrete within insulated formers made from fire-retardant expanded polystyrene. The system offers high-speed construction and low levels of air permeability.

Strength and stability

The requirement for the walls below ground to provide a sound base from which to construct the superstructure of the dwelling is perhaps the main factor affecting the form of these walls. The Building Regulations require that the load transmitted at the base of a foundation does not exceed 70 kN/m run. This restriction, together with an assessment of the safe bearing capacity of the supporting strata, dictates the area of the foundation base that can be varied. The cross-sectional area of the sub-ground walls tends to be fixed by the thickness of the walls of the superstructure. Hence there is less scope for variation in the loading conditions. The levels of pressure exerted on the walls below ground level are relatively high, since the accumulated mass of the entire structure above ground will be directed through these walls. Therefore the strength of the bricks or blocks used must be sufficient to ensure that they are capable of withstanding the loads without risk of failure by crushing. Typically, blocks or bricks with a crushing strength of $5-7$ N/mm^2 will be used.

The potentially damaging effects of lateral loading from the surrounding soil are of most concern in areas of shrinkable clay. In these situations the seasonal swelling of the clay results in increased levels of lateral pressure being exerted on the walls. Mechanisms must be introduced to ensure that the effects of this are not severe enough to cause failure of the walls.

Exclusion of moisture

There are several pathways via which groundwater may enter a building (Figure 5.3). It is essential to exclude moisture from the interior of the building to ensure that the conditions internally are appropriate. The nature of the walls below ground level is such that, although they may be wet, they must not allow moisture to pass into the building. This is achieved by the introduction of specific moisture-resisting elements. These are termed damp-proof courses (DPCs).

DPCs are placed to prevent ingress of moisture from several sources. Their position is dictated by the need to deal with rising moisture and rain splash. A variety of

Figure 5.3
Modes of water passage into buildings.

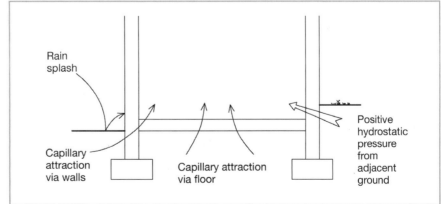

materials has been used historically, including slate and modern pitch polymer materials.

DPCs are positioned in internal and external walls. They must link with damp-proof membranes (DPMs) in floors to create a continuous barrier. They are normally placed at a minimum of 150 mm above external ground level to avoid rain splash. They also feature 'weep holes' in the brickwork beneath, to allow water from the wall cavity to escape to the exterior.

Differing ground levels

In some cases, the construction of buildings on sloping sites is necessary (Figure 5.4). In such circumstances the walls below the DPC level may be required to separate a void beneath the ground floor from the external ground. In these circumstances the wall will need to fulfil functions additional to those discussed previously. The walls extending below ground level, often to a considerable depth, are required to resist the lateral loadings that are exerted by the surrounding earth. In this respect they may be considered as retaining walls, as well as supporting the walls and super-structure above. They must, therefore, be of robust form and be capable of with-standing the range of forces applied, while remaining durable in the sub-ground environment.

In addition to the requirements for structural stability, the exclusion of ground water will generally be necessary; in some instances, however, water penetration may be acceptable to some extent, where sub-floor drainage is provided, for example. In such cases the long-term build-up of water may be prevented by pumping the water away via a sump. The exclusion of moisture from these sub-ground void areas requires the provision of vertical and horizontal damp-proofing elements around the entire sub-ground structure. This method of moisture exclusion is termed 'tanking'. Impervious materials are used to form the tanking, such as mastic asphalt, bitumen-based materials or any of an extensive range of proprietary products, often of an elastomeric nature. It is essential that tanking is continuous, as with DPCs and DPMs, and it must retain its integrity throughout the area that is to be waterproofed. Great care must be taken, therefore, where elements of the structure, such as beams and columns, together with building services, penetrate the sub-ground floor or walls. This is of particular importance, since the ground water acts against these walls with positive hydrostatic pressure. The damp-proofing elements must be capable of resisting such pressure.

REVIEW TASKS

- Outline the factors that can affect the durability of walls below ground.

- What do you understand by the term *tanking*?

- Visit the companion website at www.palgrave.com/engineering/riley1 to view sample outline answers to the review tasks.

Figure 5.4
Building on sloping sites.

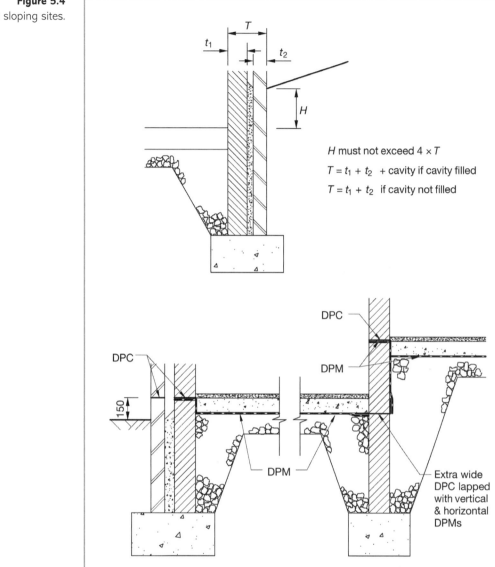

H must not exceed $4 \times T$

$T = t_1 + t_2$ + cavity if cavity filled

$T = t_1 + t_2$ if cavity not filled

5.2 | Options for walls below ground

Introduction

- After studying this section you should be familiar with the main construction alternatives available for the formation of walls below ground level.
- You should appreciate the forces applied to these sections of walls and the ways in which the construction form attempts to deal with them.

■ You should understand the implications of soil type and ground level for the choice of design solution.

■ You should appreciate the cost implications of the selection of the various options and, given specific scenarios, be able to make valid judgements regarding selection.

Overview

The requirements for walls below ground level were established in the previous section. Clearly, the main functional needs are the same for all dwellings. However there are a range of variable factors linked to the nature of the site and the soil characteristics. As with all elements of the construction of dwellings there is likely to be a range of acceptable alternatives in any given situation. Given this acceptable range of options, the basis for selection will relate strongly to cost. Generally, the most cost-effective solution will be chosen. The economy of the chosen option does not rely simply on the materials being cheaper, but also on the simplicity of the design and the level of labour input required. Speedy forms of construction will normally be simpler and cheaper because of the saving in labour costs associated with the reduction in time required.

For the construction of 'traditional' dwellings there are two main options for the construction of walls below ground: the filled cavity option and the foundation block option. These could also be utilised, with suitably amended details, for timber frame construction. Also, in some instances there is the option of adopting deep strip or trench-fill foundations. Although this is actually a foundation option and technically not a wall, its adoption largely negates the need for walls below ground level. Hence it is considered here in overview.

Filled cavity construction

The wall below ground level must provide a base for the construction of the superstructure walls. The thickness of the sub-ground wall is therefore dictated by the thickness of those above ground, rather than by specific functional criteria. The lower wall section must not be significantly narrower than the upper section (Figure 5.5).

For this reason the thickness of the wall below ground level is generally designed to be the same as that above ground. Hence, where traditional cavity construction is adopted the wall must be at the very least 250 mm thick. With changes in the thermal requirements for external walls this thickness has increased in recent years. Therefore walls of 300 mm or more are not uncommon. For the reasons outlined earlier it is undesirable to create a cavity below ground level, as the risk of the two leaves of the wall being squeezed together must be avoided. One way to achieve this is to create a solid wall section below ground. The creation of this by building a thick, bonded section would be costly and slow. Hence it is usual to create a cavity wall which then has the cavity filled to ground level with a weak mix of concrete (Figure 5.6). This achieves the desired result and is, at the same time, cost-effective.

Figure 5.5
Building Regulation
requirements for
allowable overhang.

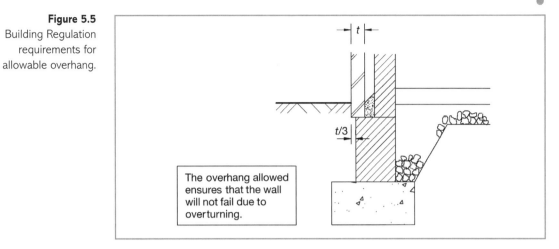

The overhang allowed
ensures that the wall
will not fail due to
overturning.

Even where the outer leaf of the wall above ground is formed in brick it is common to build the section below ground in dense concrete block. This again is for reasons of economy. Concrete blocks are cheaper and faster to lay than bricks, and since they are hidden below ground there is no aesthetic requirement.

Foundation block construction

A further efficiency gain and resultant cost saving can be made by the utilisation of larger blocks specifically designed for the purpose (Figure 5.7). Foundation blocks are available in a variety of widths to accommodate the commonly used thicknesses

Figure 5.6
Filled cavity construction.

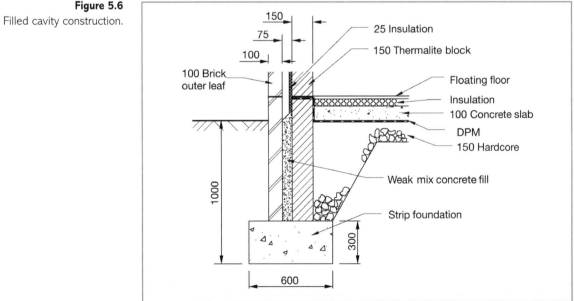

Figure 5.7
Foundation block
construction.

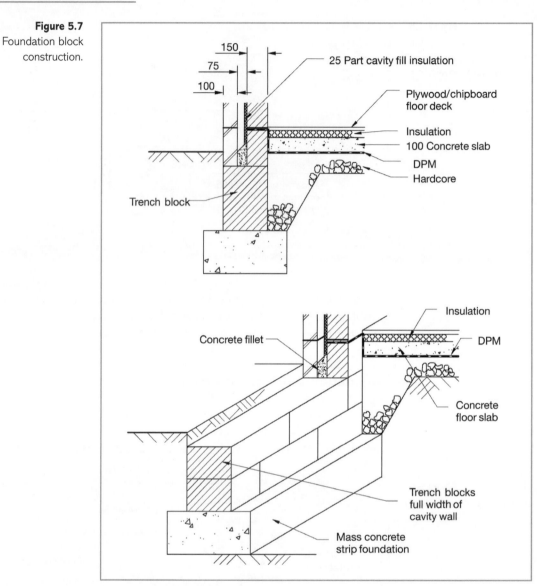

of cavity wall. The dimensions of the block allow the full thickness of the cavity wall above to be facilitated in one mass below the ground. Although this results in blocks that are considerably larger than those used above ground, and thus more difficult to manipulate, there is still an efficiency benefit. Rather than building two separate leaves of cavity walling, that then have the cavity filled with concrete, the wall is created in a single operation. In this way there is a considerable saving in time and labour. An added benefit is that the trench excavated for the formation of the foundations and walls below ground can be backfilled very shortly after construction.

Trench-fill foundations

Trench-fill foundations
are also referred to as
'deep strip' foundations.

The formation of **trench-fill foundations** is associated with soils where there is a high level of cohesion between the particles, such as clay soils. The reasons for their adoption have already been explored in detail and will not be reiterated here. However, they are worthy of mention since their use combines some of the functions of walls below ground with those of the foundation. The casting of concrete to just below ground level provides a stable base for the construction of the walls above ground and is fast and economical.

Figure 5.8 summarises the options for constructing walls below ground.

Figure 5.8
Overview of options for
walls below ground.

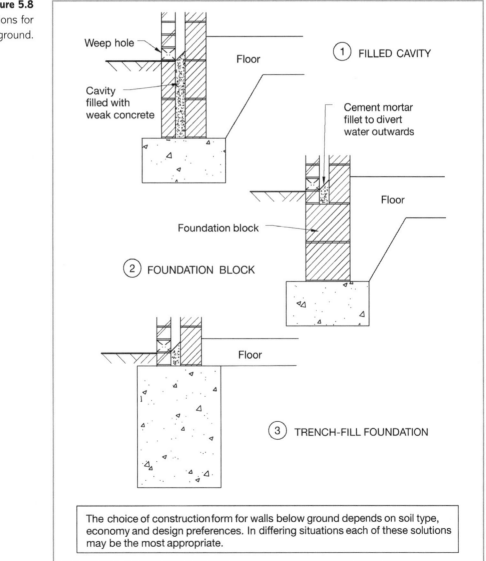

The choice of construction form for walls below ground depends on soil type, economy and design preferences. In differing situations each of these solutions may be the most appropriate.

PART 2

REVIEW TASKS

- Give *two* reasons for filling the cavity of external walls below ground level when traditional strip footing foundations are used.

- What advantages are gained by using foundation blockwork?

- Outline the advantages of deep strip compared with traditional strip foundations.

- Consider the following alternatives for walls below ground:

  - Open cavity construction
  - Filled cavity construction
  - Trench block construction
  - Deep strip foundation.

  What are the advantages and disadvantages of each option?

- Sketch each of the above options and indicate the location of DPCs.

- Visit the companion website at www.palgrave.com/engineering/riley1 to view sample outline answers to the review tasks.

COMPARATIVE STUDY: WALLS BELOW GROUND

Option	Advantages	Disadvantages	When to use
Filled cavity	– Familiar technology – Natural continuity of wall construction above and below ground – Variable width	– Need to lay bricks in trench – Several construction activities involved – Relatively costly in terms of time	– Traditional cavity construction where access to the trench is good and depth to foundation is not excessive
Foundation block	– Cheap – Single construction activity – Reduced laying time in trench	– Limited range of widths – Need to lay large units in trench	– Traditional cavity or solid wall construction where access to trench is good and depth to foundation is not excessive
Trench-fill foundation	– Cheap – No need for temporary support to trench – Saves time compared with bricklaying in trench	– Unsuitable for cohesionless soils – Need to consider services entries below ground – Potential for damage in highly shrinkable clays	– Traditional cavity or solid wall construction in cohesive soils where trench will support itself without propping and where required depth is not excessive

5.3 | Entry of services

Introduction

- After studying this section you should be aware of the main services that must be provided to dwellings.
- You should have developed an appreciation of the manner in which the various service entries must be accommodated.
- You should have developed an understanding of the factors that affect the mode and design of these various entries.
- You should have a broad awareness of governing regulations.
- You should appreciate the implications of passing services through the structure of the dwelling below ground.

Overview

The term 'services entries' is actually rather misleading. This section also deals with services that exit the building. The penetration of the external envelope of the dwelling by incoming and outgoing services is unavoidable. It is not the intention to deal with the services specifically here; however, a broad awareness of the services involved is necessary to understand the implications for the design and formation of appropriate entry and exit routes.

The mains services supplied to the building are generally referred to as **utility services** or 'utilities'. In the past these were provided by public sector providers, but more recently they have shifted to a number of private sector utilities companies. Although this is the case, they are still tightly regulated by a series of Government regulatory offices. The shift to a larger number of private sector suppliers has provided consumers with the opportunity to purchase utilities from a variety of sources. This does not affect the entry of service supplies to dwellings, since all suppliers will use the same distribution infrastructure for gas, electricity and, in most cases, telecommunications. An exception to this is the provision of cabled telecommunications, for which the infrastructure is still developing and is being supplied by cable service providers.

Building services are generally split into the categories of Utility Services and Environmental Services. **Utility services** include gas, electricity, water and drainage. Environmental services include heating, lighting, ventilation and so on. Environmental services are supported by the provision of utility services.

The nature of services supplies to dwellings

Service supplies to dwellings can be categorised in a number of ways. One of the most appropriate ways of grouping them for the purposes of this book is to consider them with reference to the form that their distribution infrastructure takes. The three forms that we are concerned with are large diameter pipes, small diameter pipes and cables. The services involved are set out in Table 5.1.

The companies responsible for the distribution of these services are also responsible for providing service connections to the dwelling. This is not necessarily as

Table 5.1 Forms of service supply.

Form	Description
Large diameter pipes	Drainage waste pipes (up to 100 mm)
Small diameter pipes	Water incoming main supply (25 mm) Gas incoming main supply (25 mm)
Cables	Electricity main supply Telephone connection Cable telecommunications link (may include TV, telephone and so on)

clear an issue as it might appear, since several companies may compete locally for service supply. In any event there will be a need to coordinate the supply of services during the construction of the building to ensure that essential work is undertaken efficiently without unnecessary duplication. In order to aid coordination of services, advice is provided by the National Joint Utilities Group (NJUG). This group is funded by the utilities providers and provides a focus for their common activities and requirements. In particular, the NJUG makes recommendations relating to common services entries, entry installation and the positions of meters.

In the past it was common for each of the service providers to dig individual trenches for their service supply to the dwelling. This led to numerous trenches being formed and to a prolonged period of installation. The advice of the NJUG is to provide a common trench for the routing of service supplies to the dwelling (Figure 5.9).

Figure 5.9
Combined services trench.

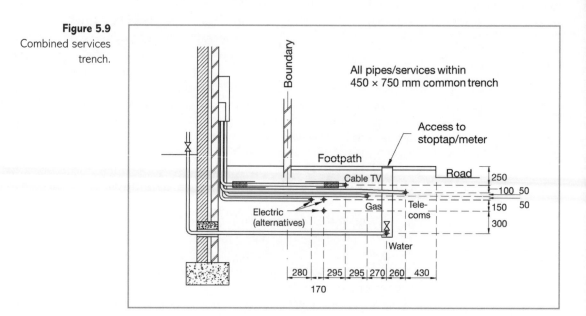

The nature of the service entries to buildings has been improved with the development of flexible materials for ducting and pipework. These allow for more continuous lengths of service pipework and ducting, with fewer joints required to cope with minor changes in direction. Threading of services around potentially troublesome obstacles is also facilitated.

It is worthwhile considering each of the main groups of services in a little more detail.

Large diameter pipes: drainage

The waste and soil pipework for sanitary appliances varies in size from 32 mm to 100 mm in diameter. It is not the intention here to consider the sanitary installation, merely to examine the implications of the need to route the associated pipework through the building fabric. The smaller sizes of waste pipework are restricted to use above ground, and passing them through the external walls is effected by simply cutting a hole of the appropriate size. There is no need for additional support to the surrounding walling.

Of more concern in relation to the walls below ground are the larger sizes of pipework used beneath the ground (Figure 5.10). Underground drainage is required within the building enclosure to connect to water closets serving the ground floor and as a result of the increased use of internal soil pipes to serve the upper floor sanitary appliances. Where these pipes pass through the external walls below ground there is potential for failure arising from differential movement. There are two major potential causes of such movement. Firstly there is the risk of ground movement resulting from frost-induced heave in the soil surrounding the building. Secondly there is the risk of movement associated with the initial settlement of the new building. All new buildings settle to a small extent owing to the minor consolidation of the supporting ground following the construction of the building. This is generally no cause for concern and will not affect the stability of the property. However, where pipes pass through walls that are likely to be affected the movement must be accommodated. If this is not so there is the risk of fracture or deformation of the drainage pipes. The consequence of this would be failure and leakage of the below-ground pipework. In order to avoid this there are two alternative approaches to the installation of the pipework. The first of these relies on the use of flexible joints to allow the pipework, which is built into the wall, to deform within acceptable limits. The second relies on the use of supporting lintels over a larger opening. The size of the opening is such that the slight movement can be accommodated without affecting the pipework. In this case, the pipework is surrounded by compressible material and a rigid patress is provided to seal the opening and prevent the entry of vermin and so on.

Small diameter pipes: water and gas supplies

The pipework providing water to the dwelling is in two parts. The section from the main supply to the meter or stop valve at the boundary of the property is termed the 'communication pipe' and is installed by the water company. The section from the boundary to the dwelling is termed the 'service pipe' and is provided by the prop-

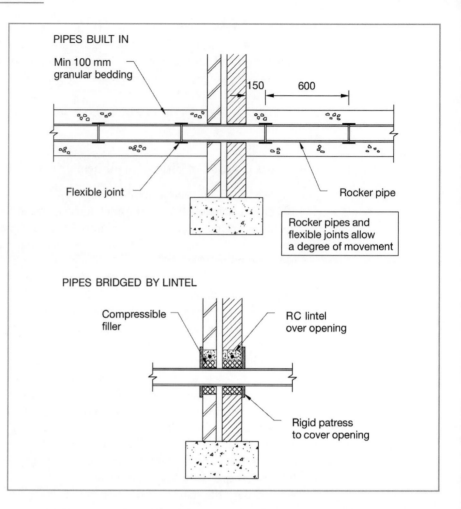

Figure 5.10

Pipework passing through walls below ground.

erty developer or the individual consumer. The service pipe will be 25 mm in diameter and will be formed from flexible MDPE pipe. The pipe is blue in colour and is easily recognisable as a water supply pipe. In order to protect the pipe from the effects of frost, and to reduce the effect of compressive loadings from vehicles and so on, it is placed in a trench of 750 mm to 1,350 mm depth. In some circumstances shallower depths are allowed, subject to the pipe being insulated against freezing. From the trench, the pipe will pass through the walls of the building below ground, rising to the stop valve inside the property. In order to provide for a degree of differential movement, the pipe is housed within a larger diameter sleeve built into the wall and passing through the floor of the dwelling. The sleeve is sealed at both ends to prevent access to vermin and to resist the passage of gas, such as methane, from the ground.

The supply of gas to dwellings is heavily regulated and the rules governing its installation are prescriptive (Figure 5.11). The service pipe is buried at least 375 mm below ground and exits the ground at the external face of the dwelling wall. In older buildings it is common to find the meter for the gas supply located within the prop-

Figure 5.11
Service entry details.

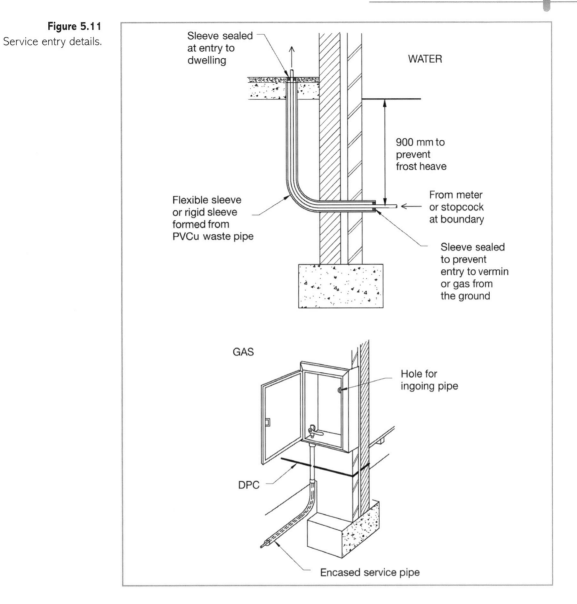

erty. In more modern buildings the meter will be located externally and is housed within a protective meter box. This box is accessible to the gas provider for the purpose of meter reading and to allow the supply to be cut off in emergencies. From the meter box the supply pipe will pass into the dwelling through the external wall above ground, and will be provided with a sleeve to prevent leakage of gas into the cavity. The details of this installation are dealt with in Chapter 2.

Cabled service: electricity and telecommunications

The nature of cabled services is such that they are far more easily accommodated than piped services and their entry to dwellings reflects this. In the case of electricity

Figure 5.12

Electrical service entry boxes.

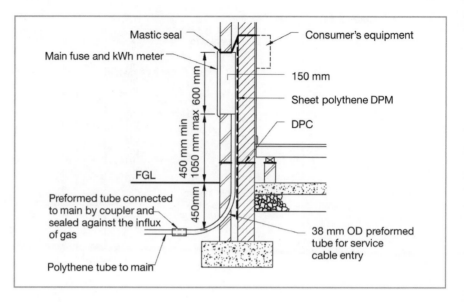

supplies, as with gas, external meter boxes are now common (Figure 5.12). There is far more freedom in positioning the cables to supply the meter and they can rise within cavities if required. The nature of the telecommunications industry is such that services are changing and expanding rapidly. Hence it is prudent to consider the use of a ducted entry within the common service trench to allow for future expansion. Once at the external wall of the property they are connected to distribution boxes externally, and their entry through the external wall takes place above ground.

REVIEW TASKS

- Name *four* service connections typically made to modern homes.

- Name the *three* forms in which these services are classed.

- Explain the reasons for the use of a common services trench.

- What are the potential risks in its use and how might they be overcome?

- Visit the companion website at www.palgrave.com/engineering/riley1 to view sample outline answers to the review tasks.

COMPARATIVE STUDY: SERVICES CONNECTIONS

Form	Option	Colour	Entry treatment	Protection
Cables	Electrical mains	Black	Entry via wall-mounted box containing meter with external access	Typically 450 mm below ground level
	Cabled services	Various, depending on provider of data – telecommunications and so on	As required by cable service provider	Consider separation of data cabling from power cables to avoid danger of electromagnetic interference
Small diameter pipes	Gas	Yellow	Entry via external meter housed in protective box with external access; ensure adequate sleeving of pipework through wall cavity	Typically 600 mm minimum below ground level
	Water	Blue	Entry through wall below ground level and ground floor to stop tap internally	Minimum 750 mm below ground level to avoid freezing and damage from loads on ground; pipework housed on curved sleeve through wall and floor
Large diameter pipes	Drainage	Various, typically terracotta/brown	Provide facility for differential movement where passing through walls below ground; use of rocker pipe or lintel required	Laid to falls in granular bedding; protect from physical damage as required

PART 2

6

Ground floors

AIMS

After studying this chapter you should be able to:

- Distinguish between the various options available for the formation of ground floors to dwellings
- Understand the functional requirements of ground floors and the criteria for the selection of alternatives
- Understand the construction detailing associated with each of the potential design solutions and you should appreciate the sequence of operations involved in their formation on-site

This chapter contains the following sections:

6.1   Functions of ground floors and selection criteria
6.2   Ground-supported floor options
6.3   Suspended floor options

INFO POINT
- BRE Good Building Guide 25 Radon: Guidance on protective measures for new buildings (1996)
- BS 6515: Specification for polyethylene damp-proof courses for masonry (1984)
- BS 8102: Code of practice for the protection of structures against water from the ground (1990)
- BS 8110: Structural use of concrete (1997)
- *Beam and block floor systems* (2005), Precast Flooring Federation
- *Precast concrete floors for housing* (2005), Precast Fooring Federation
- *Insulating ground floors* (2001), Building Research Establishment, Garston, Watford

6.1 | Functions of ground floors and selection criteria

Introduction

- After studying this section you should have developed an understanding of the functional requirements of ground floors to dwellings.
- You should appreciate the implications of ground conditions and features of the chosen site upon the selection of ground floor options.
- You should be aware of the implications of foundation and external wall construction forms upon the selection of ground floor design solutions.

Overview

Ground floors to dwellings can take a number of forms, depending, among other things, on the nature of the site, the quality of construction and the required speed of erection of the building. However, all the available design solutions essentially fulfil the same functional requirements. The detailed functional requirements will be examined in detail later within this chapter. Ground floors can be categorised into two basic groups: suspended floors and ground-supported (or solid) floors. In simple terms, the distinction between the two forms is as follows (Figure 6.1):

- Solid floors are formed such that the underside of the floor is in continuous contact with and is supported by the ground.

Figure 6.1
Ground floor
classification.

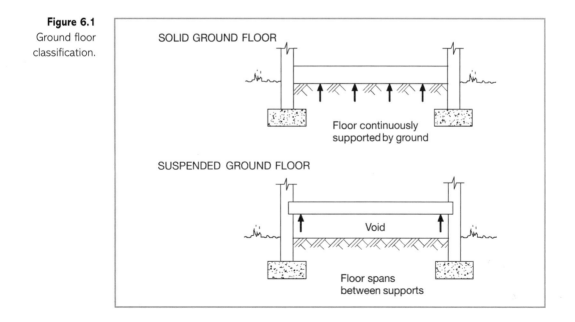

PART 2

■ Suspended floors are formed such that the structural elements of the floor span between supports, not relying on the ground for support of the floor structure. This may result in the creation of a void beneath the floor and the ground.

Functional requirements of ground floors

The main function of the ground floor is to provide a safe and stable platform for the activities that are carried out within the dwelling. However, in addition to this there are a number of other equally important functions that must be fulfilled by the floor if it is to satisfy user needs and other requirements such as the Building Regulations.

The nature of the construction form of ground floors is dictated by the relative importance of these aspects of performance. In addition, there are a series of aspects related to the selected site and the general construction form of the dwelling that will have an impact on the choice of floor; these will be examined in detail later. Let us first consider the generic performance or functional requirements for ground floors of all types. These functional requirements can be summarised as follows.

Structural stability

The ground floor of any dwelling must be designed and constructed in such a way that it is capable of supporting the dead loads and live loads that it is likely to be subjected to. Hence the construction form must be such that the floor is robust enough to resist or transfer these loads without undue deformation or the risk of structural failure (Figure 6.2). In the case of ground-supported floors this relies on the floor structure, which is in continuous contact with the ground, dissipating the loads effectively. In the case of suspended forms the mechanism is one of load transfer to the supporting elements of the structure, normally taking the form of dwarf or sleeper walls at low level.

Thermal insulation

As with all elements of the external fabric of buildings, there is a requirement for ground floors to provide a degree of resistance to the passage of heat (Figure 6.3). The degree to which this results in the need for the installation of an insulative material varies, depending on the floor structure and formation. However, as the required levels of insulation demanded by the Building Regulations have increased, so the need to provide specific insulative materials as part of the floor composition has increased accordingly. The nature of the heat loss through ground floors differs between suspended floor types and ground-supported floor types.

In the case of solid floor types, the formation of the floor in contact with the ground results in the creation of a protected zone at the core of the floor area. This effectively provides a degree of insulation to the central area of the floor, which inhibits heat loss. In periods of cold weather it can generally be assumed that the interior of the building will be warmer than the exterior, including the ground which

Figure 6.2
Load transfer through
floors.

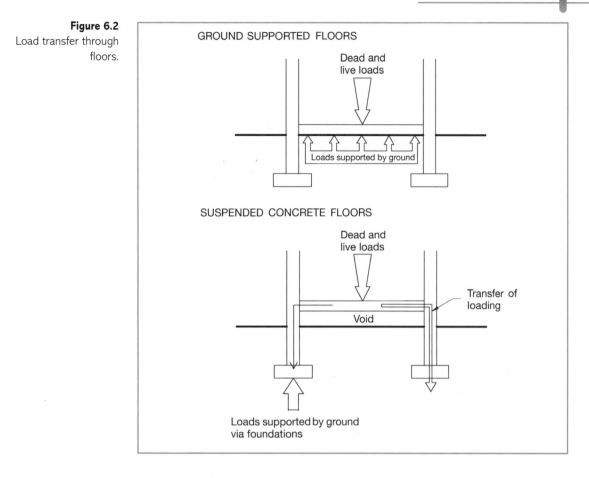

supports the building. The ground beneath the floor will itself absorb a degree of heat from the building. In addition, around the perimeter of the floor there will be heat transfer to the surrounding ground and to the external atmosphere. There will also be heat loss through the edges of the floor at the abutments with the external walls of the building close to ground level. These mechanisms will not occur at the centre of the floor, where heat loss will eventually achieve a 'steady state' and will be much reduced in extent.

For this reason the level of heat loss at the edges of the floor will be far greater than at the central area. Hence in the case of large floor areas it may be economical to insulate the floor to differing levels at different locations. In the past this has taken the form of insulation provision only at the perimeter of the floor, with the central area being left uninsulated. However, in the case of floors to dwellings, the size of the floors tends to be too small for this approach to be economical. The potential cost saving associated with materials is counteracted by the added complexity of the differing construction detailing. Hence it has generally been the case that ground-supported floors to dwellings have been fully insulated. With proposed increases in the insulation requirements for floors included within the revised Building Regulations, the amount of insulative material required will increase further.

Figure 6.3
Heat loss through floors.

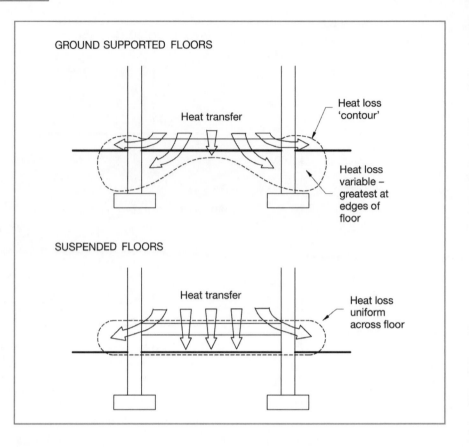

In the case of suspended floor constructions the creation of a void beneath the floor results in a more uniform level of heat transfer across the floor. The void will be subject to the passage of air, and thus the creation of a steady-state central area will not be possible. Hence it is essential to provide a uniform level of insulation across the whole of the floor. As in the case of ground-supported floors, the potential does exist for heat loss to occur at the edges of the floor where it comes into contact with the external walls. This must be taken into account when considering the insulation provision.

Exclusion of ground water and contaminants

The issue of preventing the passage of moisture to the interior of a dwelling is dealt with in detail elsewhere within this text, but it may be useful to remember that this need is the subject of Part C of the Building Regulations. It is essential to recognise that the potential for ground water to pass through the ground floor of a dwelling is a significant factor. For this reason the construction form of the floor, whether ground-supported or suspended, must include details designed to arrest the passage of moisture.

The normal mechanism of moisture ingress through ground floors is that of capillarity (Figure 6.4). Moisture is drawn into the building element as a result of the

PART 2

Figure 6.4
Moisture entry routes
through floors.

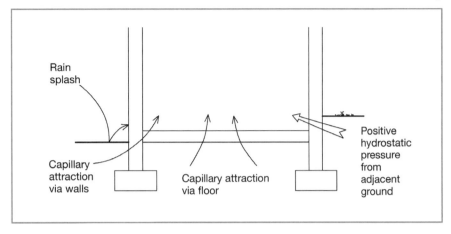

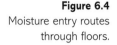

Capillary action or
capillarity is the tendency
of water to migrate
through porous materials
because of the surface
characteristics of the voids
within the material.

capillary action within porous materials such as concrete. This must be resisted by
the incorporation of impervious materials or capillary breaks within the floor. In
some instances there may be the added problem of moisture being driven into the
structure by positive pressure. This is particularly so where the level of the ground
is such that the floor structure is below the external ground level. Although such
circumstances are now rare, in older properties with cellars, or where the ground
level changes considerably, such as on sloping sites, the situation may still arise. In
these cases there is the potential for water to be forced into the building by hydrostatic pressure.

In addition to the exclusion of moisture, the floor must also be designed to ensure
that there is no risk of the passage of potential contaminants to the interior of the
building. In some areas there may be risk of ground-bourne methane or radon gas.
It is essential that these do not build up beneath the floor and, as such, specific
design details must be adopted in these areas.

Durability

As the nature of the environment within which ground floors must function is rather
aggressive, the materials that are used must be sufficiently durable to provide a satisfactory lifespan. Thus the selection of appropriate materials is essential for the
longevity of the floor.

Provision of appropriate surface finish

Ground floors to dwellings must provide a level and smooth surface finish to allow
for the provision of decorative and comfortable floor coverings. Even minor irregularities in the surface of the floor can result in premature wear of finishes. It is also
essential that the finish is level and true in terms of safety, since minor undulations
and irregularities can result in the creation of trip hazards.

As a result, concrete floors tend either to be finished with a screed (often cement
and sand-based) or power floated (see later in this chapter).

Factors affecting the selection of ground floors

A number of factors affect the selection of ground floor alternatives. These relate to the functional performance of the individual options and also take into account the following:

- The general nature of the construction form
- Nature of the site
- Anticipated loadings
- Required surface finish
- Cost.

When attempting to evaluate individual design solutions, these factors will be considered. It must always be remembered that where two alternatives are able to satisfy the functional requirements equally well, the cheaper option is almost certain to be selected.

REVIEW TASKS

- Rank the following ground floor solutions in order of your perception of cost, with the cheapest first:

  - Concrete suspended
  - Timber suspended
  - Concrete solid.

- What might influence the cost of each type?

- What may be the advantages of having a void below the ground floor?

- Why might you choose a suspended floor rather than a ground support floor?

- Visit the companion website at www.palgrave.com/engineering/riley1 to view sample outline answers to the review tasks.

6.2 | Ground-supported floor options

Introduction

- After studying this section you should have developed an appreciation of the various alternatives for the formation of ground-supported floors to dwellings.
- You should appreciate the essential elements of interaction between floors and walls.
- You should be familiar with the components of the various floor formations.
- You should understand their functions and the mechanisms by which they satisfy them.

Overview

As outlined previously, the essential characteristic of ground-supported floors is that they are in continuous contact with the ground beneath and they transfer their loads through this contact area. They are often termed 'solid floors', since there are no void or hollow areas within the construction. Ground-supported floors may take a number of forms, although all are similar in principle, differing only in detail of design and construction. All the floor options must fulfil the same performance requirements, and as such the level of flexibility in generic form is limited. However, several alternatives exist as a result of the ability to adopt different positions for individual components of the floor assembly.

Solid floor construction

As previously noted, the term 'solid floor' is often used to describe ground-supported floors; all of the options adopt the same approach to construction with the use of a layered form incorporating several individual elements. The main load-supporting element is the floor slab which will normally take the form of a layer of mass concrete, cast *in situ* to the desired level. In floors which are to take high loadings, or where the loadbearing capacity of the ground is low, the slab may be reinforced with mild steel bars or mesh. In normal circumstances a slab thickness of 100–150 mm would be used.

The surface of the floor must usually be suitable to accept a surface finish or to be trafficked directly; hence the raw surface of a cast slab is often unsuitable. The final finish may be provided in a number of ways, but it is most common to provide a layer of sand and cement screed, approximately 50 mm thick, over the slab to provide a wearing surface. An alternative to this is to use a layer of mastic asphalt, which has the advantage of being impermeable, or to grind a level finish directly on the surface of the setting slab using a power float.

As was previously noted, the exclusion of moisture is of paramount importance in areas such as ground floors; hence a damp-proof membrane (DPM) is installed above or below the slab and linked with the damp-proof course (DPC) in the walls to form a continuous barrier. The exclusion of moisture is aided also by the provision of a layer of **hardcore** beneath the floor slab. The hardcore bed consists of a layer of crushed stone, or clean, broken brick at least 150 mm thick. The layer acts to provide a level uniform base onto which the slab can be laid. In addition, however, the voids between pieces of hardcore act to break the capillary path of moisture rising from the earth. If not properly compacted, the hardcore layer may be subject to consolidation following loading of the floor; hence it is laid in thin layers about 100–150 mm thick and is carefully compacted as work proceeds. A thin layer of sand or ash is then laid over the hardcore in order to prevent puncturing of the DPM by the sharp points of the hardcore. This is termed 'blinding'. In instances where the DPM is placed above the slab, the provision of the blinding layer also acts to resist the passage of fine cement particles from the slab to the hardcore, which would weaken the finished floor.

Hardcore is a layer of crushed stone which is laid and compacted to form a base layer for the floor. The material used must be carefully selected to avoid introducing contaminants into the construction. Broken brick was once commonly used but this is now rare.

PART 2

Figure 6.5
Location of the DPM.

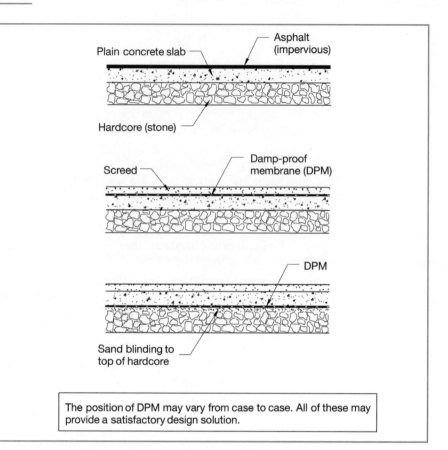

The position of DPM may vary from case to case. All of these may provide a satisfactory design solution.

The DPM may be located in various positions when constructing the slab and may even be in the form of an impervious finish such as asphalt, as shown in Figure 6.5.

When making a floor slab watertight it is essential to have an overlap between the floor DPM and the wall DPC (Figure 6.6).

The provision of insulation in the floor is necessary to reduce heat loss to acceptable levels. Typically, U (thermal transmittance coefficient) values of 0.25 W/m² K are required for ground floors under the Building Regulations. The inclusion of a layer of high-efficiency insulation material, such as expanded polystyrene, is necessary to achieve such a level of thermal resistance, and this may be placed in a number of alternative positions as illustrated in Figure 6.7.

The majority of heat loss occurs at the perimeter of the floor, often by **cold bridging** at the point of contact between the floor and the external wall. For this reason, insulation is also provided vertically at the junction of the two elements (Figure 6.8).

If a screed is to be used to finish a concrete floor slab rather than power float, there are a number of ways in which the screed itself may be laid. Figure 6.9 shows a number of alternatives; the choice depends on the nature of the building use and the likely loads in addition to the type of finish selected. Where insulation is laid over the slab and beneath the floor screed, a thicker screed is required to resist cracking, as the insulation will give under applied loads.

Cold bridging is the term used to describe the situation when the level of insulation of the fabric is significantly lower at a localised position than in the remainder of the fabric. In this circumstance heat loss is concentrated in the area of lower insulative value causing a 'cold bridge' between the warm inner and colder outer air.

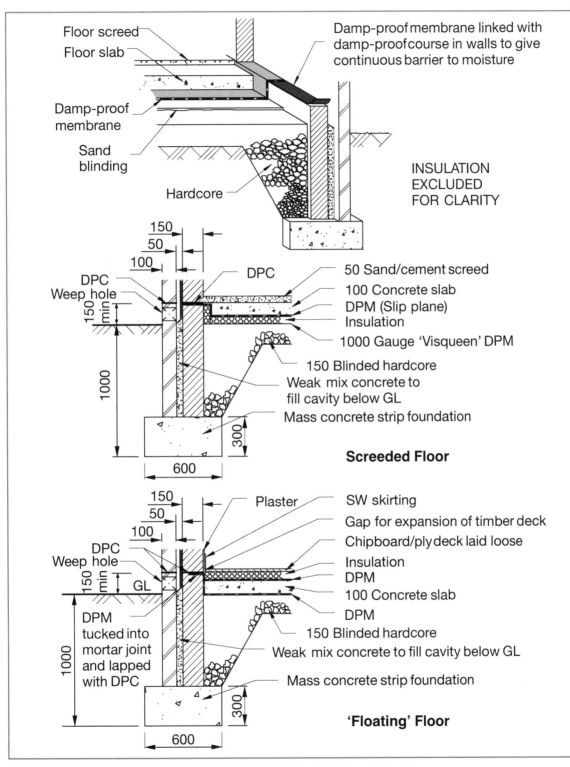

Floor screed
Floor slab
Damp-proof membrane
Sand blinding
Hardcore

Damp-proof membrane linked with damp-proof course in walls to give continuous barrier to moisture

INSULATION EXCLUDED FOR CLARITY

150
50
100

DPC
Weep hole
150 min
1000

DPC

50 Sand/cement screed
100 Concrete slab
DPM (Slip plane)
Insulation
1000 Gauge 'Visqueen' DPM

150 Blinded hardcore
Weak mix concrete to fill cavity below GL
Mass concrete strip foundation

300
600

Screeded Floor

150
50
100

Plaster

SW skirting
Gap for expansion of timber deck
Chipboard/ply deck laid loose
Insulation
DPM
100 Concrete slab
DPM
150 Blinded hardcore

DPC
Weep hole
150 min
GL

DPM tucked into mortar joint and lapped with DPC
1000

Weak mix concrete to fill cavity below GL

Mass concrete strip foundation

300
600

'Floating' Floor

PART 2

Figure 6.6 Meeting between the floor and wall.

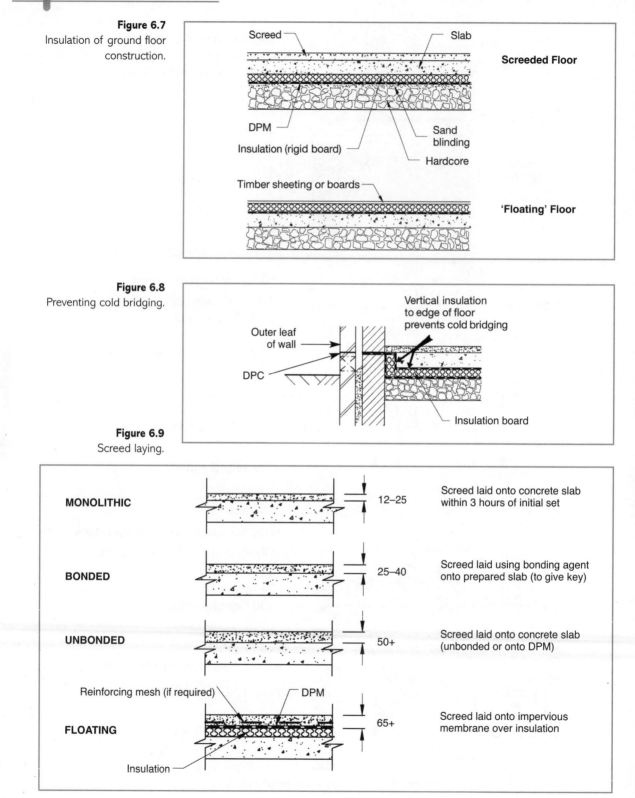

Figure 6.7
Insulation of ground floor construction.

Figure 6.8
Preventing cold bridging.

Figure 6.9
Screed laying.

Figure 6.10
Disabled access and level thresholds.

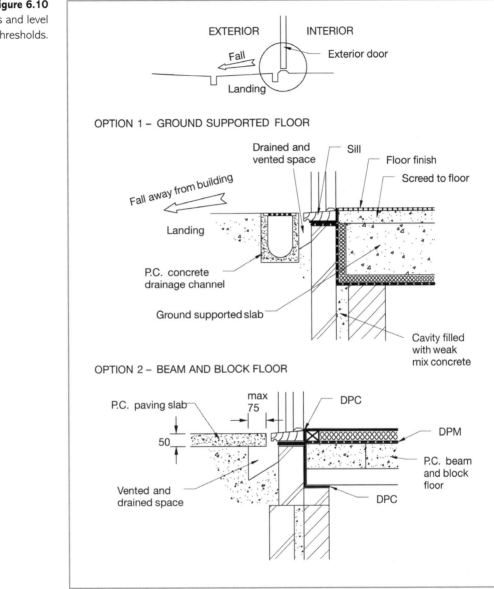

EXTERIOR   INTERIOR

Fall
Exterior door

Landing

OPTION 1 – GROUND SUPPORTED FLOOR

Drained and vented space   Sill
Floor finish
Screed to floor

Fall away from building

Landing

P.C. concrete drainage channel

Ground supported slab

Cavity filled with weak mix concrete

OPTION 2 – BEAM AND BLOCK FLOOR

P.C. paving slab   max 75   DPC

50
DPM

P.C. beam and block floor

Vented and drained space   DPC

PART 2

It is sometimes the case that special provisions need to be made with floors for situations such as disabled access. Here it is important to eliminate any difference in level between the outer paving and the internal floor level to allow a smooth transition from outside to inside. Compatible external and internal levels can be achieved at the entrance door, but creating level surfaces at this position necessitates the incorporation of details to intercept rainwater. Figures 6.10 and 6.11 indicate potential design options aimed at facilitating level access to buildings without the need for raised thresholds.

Figure 6.11
Level threshold example during construction.

6.3 Suspended floor options

Introduction

■ After studying this section you should be familiar with the various forms of suspended floor available for use in the construction of dwellings.

■ You should understand the benefits of adopting suspended floor constructions and the reasons underlying the decision to use them in dwellings.

■ You should understand the nature of the timber and concrete alternatives.

■ You should have developed knowledge of the construction details for each.

Overview

'**Wet trades**' are those construction operations that traditionally involve the use of 'wet' materials such as cement, mortar, plaster and concrete. Increasingly, 'dry' alternatives are being used to remove the time loss required while these elements dry or cure, allowing subsequent operations to take place.

In some circumstances the utilisation of ground-supported floors is unsuitable as a result of functional factors. Increasingly the wish to exclude '**wet trades**' from the construction process is becoming a factor in the selection process. For this and other reasons, the adoption of suspended floors is increasing in popularity. This is particularly the case with suspended concrete forms, although timber variants are becoming less popular, primarily because of cost and issues associated with durability.

In circumstances where it is necessary to provide a large void beneath the ground floor, the use of a suspended floor system is necessary. In its most traditional form this would have consisted of a series of timber joists, clad with a durable wearing surface such as softwood boarding. This system is still common in the construction of dwellings, but has been replaced in commercial and industrial buildings by more modern systems, capable of carrying great loads and without being subject to decay over time. Such systems are now generally based on the use of precast reinforced concrete units, although the use of simple reinforced concrete slabs, cast *in situ*, is also still an option.

Timber ground floors

Timber ground floors are also referred to as 'suspended' or 'hollow' floors. They represent the traditional form of ground floor and have many advantages to the occupier despite the high cost, which discourages builders from using this floor form. Occupiers would see these floors as aesthetically pleasing, and a modern trend is to have stained or varnished floorboards as the floor finish in preference to carpet. In contrast with concrete floors, these are also warmer and provide a degree of flexibility.

Despite their relatively high cost they have some advantages for the builder too, in that there will always be an air void below the floor, which is an extremely useful route for the installation of water pipes, such as central heating pipes, and for routing electrical cabling. Both of these services are significantly easier to install in properties with suspended floors.

The hygroscopic nature of timber results in the potential for moisture-related decay and deterioration, consequently we need to keep timbers away from any moisture. During the earlier part of the previous century it was still the practice to install the floor joists of ground floors directly into the external wall of the property, and whether the external walls were cavity or solid this laid the joists open to moisture absorption. Many of the properties of these times which still exist have had problems with their floors, and many have had their timber suspended floors replaced with a solid concrete alternative because of rot problems.

Irrespective of the age of the property it has always been the practice to ventilate the below-floor void to discourage damp conditions which might cause rot. Figure 6.12 shows this practice in a more modern timber hollow floor, where it is now standard procedure to prevent contact between the external wall of the property by standing the floor joists on supportive sleeper walls.

Figure 6.12 also shows the desired cross-ventilation supplied through the use of airbricks. It also shows the provision of timber wall plates which take the point loads from each joist and spread them along the supporting sleeper walls. These walls are built honeycombed (with bricks missing) to allow the air below the floor to circulate adequately. One of the reasons that this floor is so relatively expensive compared with other alternatives is that effectively a solid concrete floor is needed to support the sleeper walls. This concrete, as shown, is generally referred to as the 'oversite'

Figure 6.12
Timber suspended floor
details.

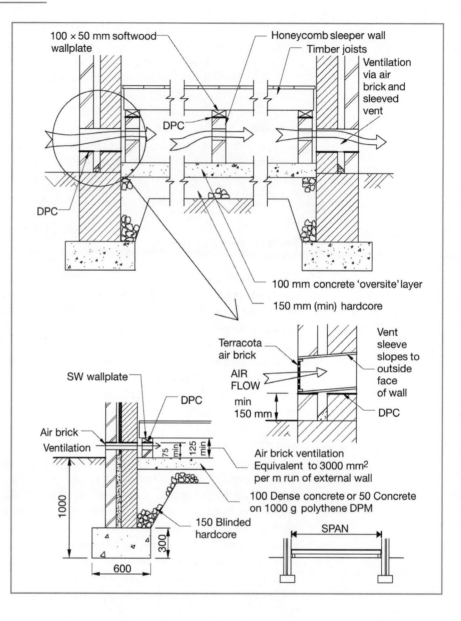

Figure 6.12
Timber suspended floor details.

concrete layer and this, as with a concrete floor slab, requires an adequate layer of hardcore for support.

Figure 6.13 shows the same timber floor in a projected section, helping to show the make-up of the sleeper wall and the function of the wall plate in greater detail.

One criticism that might be made of the timber suspended floor is in relation to heat conservation. There is only the floorboarding between the inside of the premises and the air void below the floor. The temperature below the floor is generally low, and this, with the air circulation needed, helps preserve the timber; but of course this also encourages heat loss through the floor. This can be quite easily combated by the provision of rigid insulation between the joists, as illustrated in Figure 6.14.

Figure 6.13
A timber suspended floor
section.

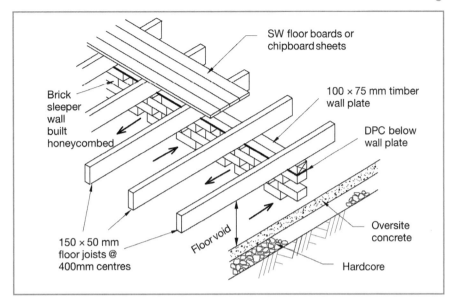

Figure 6.13
A timber suspended floor
section.

This figure also shows the edge detailing needed if cold bridging at the floor edge is to be avoided.

Beam and block floors

The formation of beam and block type floors relies on the provision of a series of profiled reinforced beams, set at relatively close spacings (typically 600 mm), spanning between supporting walls. The beams are often an inverted T-shape to allow support to be given to the infill blocks that are dropped in between the beams. A sand/cement screed is then applied over the floor to give a smooth surface finish to receive carpet and other finishes.

This system has the advantage that it comprises a series of small units which can be readily manipulated on-site, without the aid of mechanised plant.

As shown in Figure 6.15, it is acceptable to support this type of floor on the inner skin of a cavity external wall. This is of course because this floor is not in timber but concrete. Where the floor is supported by walls, whether external or internal, a DPC will always be located below the floor units, as illustrated.

Infill blocks are dropped between the reinforced concrete beams and it is easy to finish off the blocks at the end of the beams by cutting blocks to length. At the other floor edge, though, there tends to be a small gap between the beam just inside the external wall and the wall itself. Figure 6.16 shows two ways of supporting the finishing floor screed at this location.

This figure also shows that, provided topsoil is removed, there may be no need to supply oversite concrete with this type of floor. Additionally, as the floor is not liable to perish owing to dampness in the atmosphere of the below-floor void, there is no need to supply ventilating airbricks. This also means that there is less likelihood of heat loss with this floor compared with the timber alternative. It is, however, possible

Figure 6.14
Insulation of timber
suspended floors.

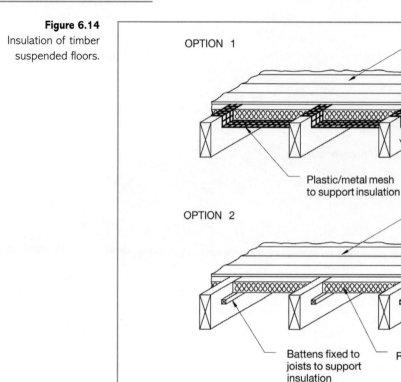

OPTION 1

Flooring deck

Insulation

Timber joists

Plastic/metal mesh
to support insulation

OPTION 2

Flooring deck

Battens fixed to
joists to support
insulation

Rigid insulation material

EDGE DETAIL
To avoid cold
bridging

to include insulation in the beam and block floor to meet the insulation requirements of the Building Regulations (Figure 6.17).

If support needs to be provided to internal partition walls, it is possible to use double beams and wet concrete infill below the partition, as shown in Figure 6.18. This detailing effectively creates a substantially larger reinforced concrete beam below the partition.

Some alternative details are given in Figure 6.19 in a summary of some of the different floor options available with the beam and block floor solution.

Figure 6.15
Suspended precast concrete (beam and block) flooring systems.

Figure 6.16
Edge detailing – beam and block floors.

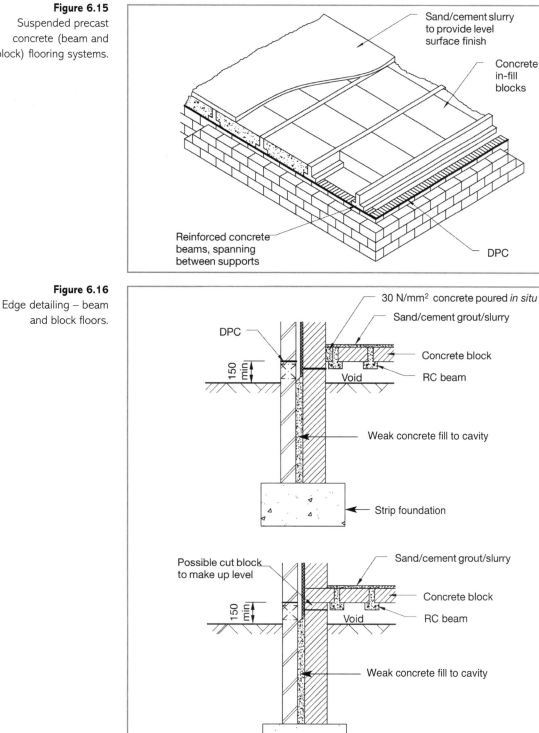

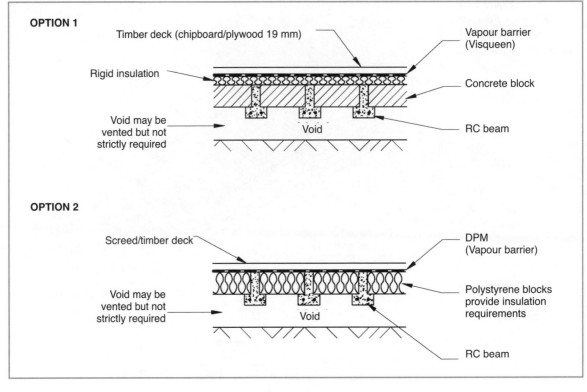

OPTION 1

Timber deck (chipboard/plywood 19 mm)

Vapour barrier
(Visqueen)

Rigid insulation

Concrete block

Void may be
vented but not
strictly required

Void

RC beam

OPTION 2

Screed/timber deck

DPM
(Vapour barrier)

Void may be
vented but not
strictly required

Void

Polystyrene blocks
provide insulation
requirements

RC beam

Figure 6.17 Insulating beam and block floors.

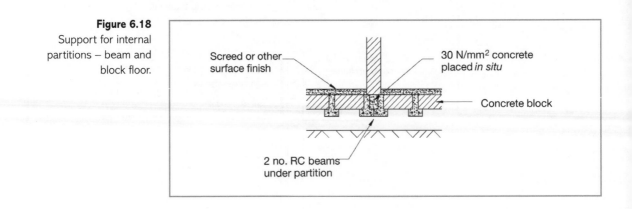

Figure 6.18
Support for internal
partitions – beam and
block floor.

Screed or other
surface finish

30 N/mm² concrete
placed *in situ*

Concrete block

2 no. RC beams
under partition

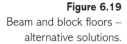

Figure 6.19
Beam and block floors –
alternative solutions.

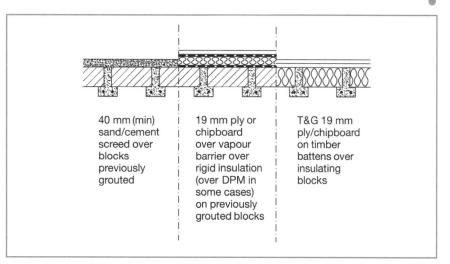

40 mm (min)
sand/cement
screed over
blocks
previously
grouted

19 mm ply or
chipboard
over vapour
barrier over
rigid insulation
(over DPM in
some cases)
on previously
grouted blocks

T&G 19 mm
ply/chipboard
on timber
battens over
insulating
blocks

REVIEW TASKS

■ How can the insulating qualities of the traditional timber suspended floor be improved?

■ Give *two* reasons for using sleeper walls.

■ How is a beam and block floor finished to receive homeowner-applied floor finishes such as carpet?

■ What advantages do beam and block floors have over the timber alternative?

■ What are the main features that must be considered when creating a level threshold detail to the main entrance of a dwelling?

■ Visit the companion website at www.palgrave.com/engineering/riley1 to view sample outline answers to the review tasks.

PART 2

COMPARATIVE STUDY: GROUND FLOORS

Option	Advantages	Disadvantages	When to use
Solid concrete	– Cheap – Familiar technology – Easy detailing to insulate and resist moisture – Good loadbearing performance	– Wet operation slows construction – Several operations involved – Expensive when dealing with deep voids	– Most common form of ground floor – Used on relatively level sites with traditional foundation design
Suspended timber	– Resilient floor form – Can cope with sloping sites and deep floor voids	– Potential problems of decay if subject to moisture – Expensive – Limited loadbearing capacity	– Rarely used in modern house building – Sometimes adopted for high-quality projects
Beam and block	– Cheap – Fast – Good loadbearing capacity – Copes with sloping sites and deep voids	– More expensive than solid floor construction on level sites with small voids	– Becoming very common in house building – Particularly appropriate for sloping sites, sites with poor ground conditions and sites involving deep sub-floor voids

Building superstructure

7

External walls

AIMS

After studying this chapter you should be able to:

- Appreciate the functional performance characteristics required of external walls
- Relate functional performance to the design alternatives for external walls
- Understand the relationship between the changing thermal insulation requirements for walls and the evolution of wall design
- Appreciate the details needed to ensure wall stability
- Understand the role played by the wall in the transfer of loads from other elements, such as upper floors and the roof

This chapter contains the following sections:

INFO POINT

- Building Regulations Approved Document A, Structure (2004 including 2010 amendments)
- Building Regulations Approved Document L, Conservation of fuel and power (2010)
- BS EN 845: Specification for ancillary components for masonry. Lintels (2003)
- BS 1243: Specification for metal ties for cavity wall construction [no longer current but cited in Building Regulations] (1978)
- BS 3921: Specification for clay bricks (1985)
- BS 5268: Structural use of timber (2006)
- BS 5977: Lintels (1981)
- BS 6073: Precast concrete masonry units. Specification for precast concrete masonry units (1981)
- BS 8103: Code of practice for masonry walls in housing (2005)
- BS 8215: Code of practice for design and installation of damp-proof courses in masonry construction (1991)
- NF41: *Low and Zero Carbon Homes: understanding the performance challenge* (2012) NHBC Foundation
- P 301: Building design using cold formed steel sections: light steel framing in residential construction (2001)

7.1 | Functions of external walls and selection criteria

Introduction

- After studying this section you should be aware of the functional requirements of external walls.
- You should be able to understand their implications for construction form.
- You should appreciate the basis of each of these and the drivers behind the development of increasing levels of performance requirement.
- You should have developed an appreciation of the overlapping of certain functional requirements and be able to make judgements about their relative importance.
- Given specific scenarios, you should be able to set out a series of criteria for the selection of appropriate external wall solutions.

Overview

The performance requirements of the building fabric have been discussed previously; however, it is appropriate to summarise them here as they relate directly to the construction of external walls. In general, they may be considered to include the following:

- Strength and stability
- Exclusion of moisture/weather protection
- Thermal insulation
- Durability
- Acoustic insulation
- Aesthetics.

Additionally, the level of buildability is important. 'Buildability' is the term used to provide a measure of the complexity of the building form and the construction detailing. This is increasingly important, since it has a direct effect on time, cost and quality in construction.

When considering any one of the potential design solutions for external walls, it is important to remember that the end result is a consequence of the need to satisfy these functional requirements.

Strength and stability

The achievement of required levels of structural stability is essential if the building is to withstand the loads that are imposed upon it during its life. Vertical, oblique and lateral loadings must be safely transmitted through the structure to the loadbearing strata. The external walls may or may not take an active role in this transmission, depending on the structural form of the building. In some situations the external walls act only as weatherproofing for the building, carrying none of the dead or live loads from the structure.

It is possible to consider external walls in categories related to the extent to which they act as loadbearing elements of the building.

Loadbearing walls

In modern domestic construction, and in many older buildings, most external walls are designed to be loadbearing walls (Figure 7.1), in that they carry their own self-weight, together with some element of loading from the rest of the building, such as floor or roof loadings. Masonry is the most common material utilised, with brick or concrete blockwork being almost ubiquitous in this form of construction. The loadings transmitted through the wall are transferred to foundations below ground level. In order to ensure that stability is maintained, certain restrictions are made within the Building Regulations relating to the height and thickness of walls and the number and positions of openings. The provision of lateral restraint is essential to resist the lateral loadings applied to buildings from wind and so on. These issues will be explored in detail later within this chapter.

Figure 7.1
Loadbearing walls.

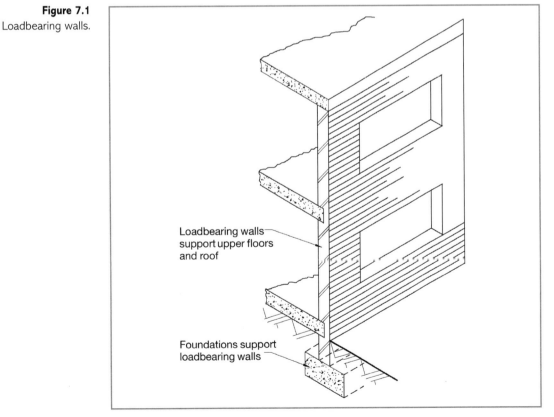

Loadbearing walls support upper floors and roof

Foundations support loadbearing walls

In the construction of low-rise dwellings the use of loadbearing masonry walls is very much the norm. This is often referred to as 'traditional' construction. In recent years, timber frame construction has also been adopted by some house builders, although its popularity has suffered because of poor publicity relating to some of the early failings of this form. For most dwellings the traditional loadbearing masonry wall and the timber frame construction are the only realistic options. One important aspect of loadbearing external house walls for the student to grasp is that, irrespective of the form of the cavity wall (masonry or timber framed), it is only the inner skin which carries structural load from the upper floor and roof.

REVIEW TASKS

■ List *three* performance criteria that we would expect to be provided by a load-bearing external wall and rank these in importance.

■ How is your most important criterion satisfied by the design of the wall?

■ Visit the companion website at www.palgrave.com/engineering/riley1 to view sample outline answers to the review tasks.

Exclusion of moisture

In the opinion of most building users this may be the main purpose of external wall construction. The ability to exclude wind, rain, snow and excessive heat or glare from the sun is paramount in the list of user requirements. Yet this must be achieved while still allowing best use to be made of natural light and ventilation. The users must, of course, also be able to enter and leave the building, thus creating the need for numerous openings to be formed in the building enclosure. These openings must be treated carefully to ensure that they do not provide a route for moisture entry to the interior of the building. In modern construction three differing approaches may be taken to achieving the exclusion of moisture, the choice depending on the nature of the building, its use and location. The three forms are solid porous construction (as used in older brick built properties), impervious cladding and masonry cavity construction (as used in most modern houses) (Figure 7.2).

Porous solid construction

External walls formed of porous materials, such as solid masonry walls, have been used with varying levels of effectiveness for many years. Although currently out of favour in England, this form of building is still very common in Scotland and other areas where high levels of exposure are experienced. Provided that the wall thickness is sufficient for the specific situation, good levels of weatherproofing are possible. This relies to some extent on the fact that long periods of continuous rainfall are rare; hence the saturated section of the wall is limited to the outer zone, as drying caused by air movement arrests the migration of water prior to it entering the interior of the building. The central zone of the wall may be almost constantly wet, since the drying effect of the air movement is unlikely to dry the full wall thickness before

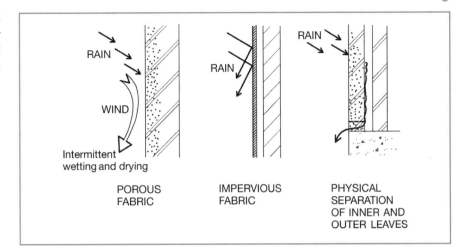

Figure 7.2
External walls –
moisture-resisting
options.

it is once again subjected to rainfall. In order to upgrade the performance of walls of this type, an external coating of render which offers a greater level of weather resistance may be used. This, however, may pose problems in the long term, with the possibility of thermal movement causing cracking, allowing water to enter and become trapped behind the protective coat.

Impervious or weather-resistant cladding

An alternative approach to the exclusion of moisture in external walls is to provide an external covering or cladding to the wall, which offers a great level of weather resistance. Materials such as steel and aluminium sheeting or panelling, glass or plastics may be used to form a continuous impervious covering to the structural wall. This form of construction is common in larger buildings, such as industrial units and multi-storey blocks, with large pane units being lifted into place with the aid of a crane. The necessity to adopt such techniques tends to make these forms unsuitable for small-scale domestic construction, which is still labour-intensive. Weather-resistant claddings may also be used for housing, such as roof tiles on felt and battens. The larger cladding sheets and panels are aesthetically unpleasant for such small-scale buildings and are restricted to larger building types, such as offices.

One potential disadvantage of this form of building is that in the event of a localised failure of the material, the level of water penetration to the interior may be great since the surrounding material does not absorb any of the water. In a porous construction form moisture is absorbed from any crack or fissure by the material itself, thus limiting the extent of passage to the interior.

Cavity construction

The most common form of external wall in use today for domestic and small-scale construction is the masonry cavity wall. The principle upon which these walls operate is to create a break in the capillary path of the moisture. Water penetration

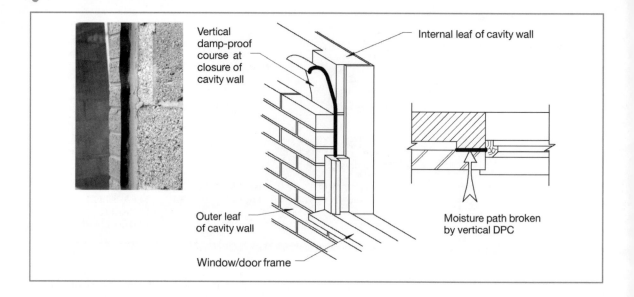

Figure 7.3
Exclusion of moisture at the cavity closure.

through porous materials relies on **capillarity**. Once the ability to move by this mechanism is removed, moisture will cease to travel towards the building interior. The outer leaf of the construction is allowed to be saturated to its full thickness; the inner leaf, however, should remain dry. Problems occur in this form of construction where the cavity must be closed for structural purposes and to resist the passage of fire. At these points the inner and outer leaves may make contact and allow the passage of heat and moisture if not treated properly. Hence details must be developed to prevent this, with insulation materials and vertical damp-proof courses incorporated in the structure (Figure 7.3).

The extent to which the passage of moisture will affect the chosen design is, to some extent, variable. The location and orientation of the building will have a considerable impact on the final choice of wall construction. In the UK it is possible to assess the variation in the degree of exposure at different locations by referring to maps that indicate the driving rain index. Clearly, the greater the level of exposure the greater the need to incorporate design features to prevent moisture penetration.

Capillarity is the tendency of water to migrate through porous materials because of the surface characteristics of the voids within the material.

REVIEW TASKS

- Give *two* reasons for the lack of popularity of solid porous wall construction.

- Visit the companion website at www.palgrave.com/engineering/riley1 to view sample outline answers to the review tasks.

Thermal insulation

'Sustainability' in the context of construction relates to the concept of constructing and using buildings in such a way that they have the minimum detrimental effect on the environment. Increasing standards of thermal insulation reflect this growing agenda.

The increasing requirements for the conservation of fuel and energy, resulting from a growing drive towards **'sustainability'**, the wishes of building users and the requirements of the Building Regulations, demand that external walls are

Figure 7.4

Insulation channels –
cavity walls.

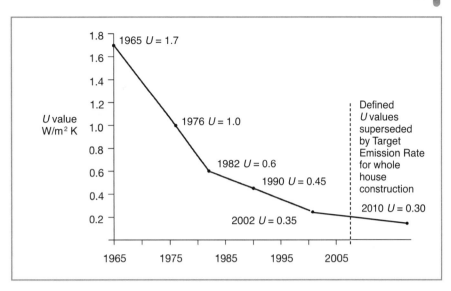

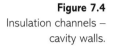

constructed to reduce heat loss to acceptable levels. Figure 7.4 shows how the required level of insulation has increased in recent years and how it is anticipated that it will increase further. All new dwellings must be provided with an 'SAP' rating calculated in accordance with the Government-approved Standard Assessment Procedure (SAP). The rating is given on a scale of 1–100 and is based on the calculated annual energy cost for space and water heating. Higher SAP ratings indicate better thermal efficiency, and although there is no requirement to achieve any particular level, buildings with ratings below 60 are considered to require higher levels of insulation. The notion of defined U values for building elements has largely been superseded by Target Emission Rate calculations as discussed in earlier sections. However, minimum elemental fabric standards are still referred to. In the most recent Building Regulations, the limiting U value of walls is 0.3 W/m^{2}K. Naturally, as the efficiency of individual elemental insulation improves there is an issue of diminishing marginal returns. At this point, issues such as air permeability take on increasing relevance. This is within the context of the overall target transmission rate.

The required levels of insulation can be achieved in a variety of ways, depending on the nature of the construction used. It has become impractical, however, to attempt to achieve this level without the aid of an insulation material of high efficiency. High-efficiency thermal concrete block is available, although the thickness which is required to meet the requirements is too great to be practical. Hence more efficient materials, such as polyisocyanurate, are used to raise insulation to acceptable standards. With the anticipated increase in required levels of insulation it is felt that timber frame or steel frame housing may become the most viable method of construction, as they are ideally suited to the incorporation of insulative materials. In the medium and long term it is possible that traditional brick/block cavity walls will become uneconomic and impractical in attempting to meet these standards. One reason for this is the limitation on the width of cavity for structural reasons.

PART 3

Figure 7.5
Temperature profiles in
walls with differing
insulation positions.

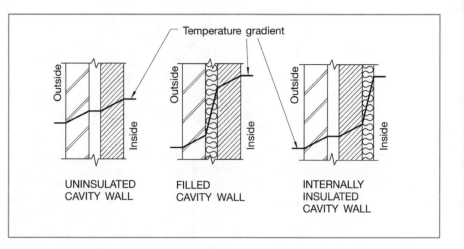

This, in turn, limits the extent to which the wall can be insulated. These considerations support the current drive towards alternative construction forms.

The positioning of insulation material can have a great effect on the thermal performance of the wall in both solid and cavity forms (Figure 7.5). The external positioning results in the wall fabric absorbing heat from the interior space. This may take some time before reaching steady state. If it is positioned internally, however, the fabric is insulated from the interior space and hence does not absorb heat; this allows the building to respond quickly to heat input. The concept of fast and slow response times or thermally heavy and thermally light construction was dealt with in Chapter 1.

In the case of cavity walls, the need to **close cavities** around openings presents potential problems in terms of insulation. This is because the bringing together of inner and outer leaves provides a direct route for the passage of heat. This is known as **cold bridging** or thermal bridging, and has been discussed earlier. Cold bridging must be avoided by the installation of an insulative material at the potential bridge points (Figure 7.6). Alternatively, a thermal break can be incorporated by using an insulative cavity closer.

It is normal practice to **close cavities** around window and door openings and at the eaves and verge of the roof. This is primarily to prevent the passage of fire through the open cavity.

Cold bridging is the term used to describe the situation when the level of insulation of the fabric is significantly lower at a localised position than in the remainder of the fabric. In this circumstance, heat loss is concentrated in the area of lower insulative value, causing a 'cold bridge' between the warm inner and colder outer air.

Durability

The life expectancy of the building and its elements depends largely on the ability of the materials used to withstand the ravages of time and the elements. In the case of external walls they are often subjected to a hostile environment, which is liable to variation between extremes of heat and cold, wet and dry and so on. Also, they must be resistant to physical damage from impact and general wear and tear, while being capable of resisting the loads imposed upon them.

In particular, the effects of fire upon durability can be considerable and have consequent effects upon other building elements and other buildings. The external walls of a building are required to contain a fire for a prescribed period of time to inhibit spread to adjacent buildings. In the case of residential buildings this period is 30

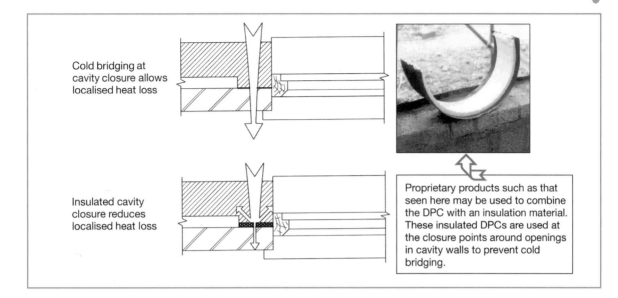

Cold bridging at cavity closure allows localised heat loss

Insulated cavity closure reduces localised heat loss

Proprietary products such as that seen here may be used to combine the DPC with an insulation material. These insulated DPCs are used at the closure points around openings in cavity walls to prevent cold bridging.

Figure 7.6

Cold bridging avoided by insulation at closure points.

minutes, although for industrial and commercial building forms this may increase to up to 4 hours, as in the case of storage buildings. The walls must also retain their integrity and stability until the occupants have had time to escape. The choice of materials dictates the performance of the wall. Performance in the event of fire is prescribed by the Building Regulations in two ways. Firstly, it deals with the ability of the fire to spread across the face of the wall. This is termed *surface spread of flame* and is rated from 0 to 4, where 0 is least flammable and 4 is most flammable. Secondly, combustibility is taken into account. This has no units of measurement, but materials are described as being combustible, non-combustible and of limited combustibility.

The choice of components that make up the wall should be undertaken with great care. Masonry walls, because of their inherent qualities, are very durable and generally do not need special protective design features or treatments. However, if materials are of poor quality or used in inappropriate combinations, premature failure may result.

Acoustic insulation

The need to minimise the level of sound transmission through external walls can arise for a variety of reasons, but in general could be considered necessary when sound levels differ greatly from the inside to the outside of a building. Sound travels in two distinct ways: via a solid material (impact or flanking sound transmission) or via the air (airborne sound transmission). Airborne sound transmission requires a massive or dense construction to reduce it, while physical breaks in the structure may stop impact sound. The creation of physical breaks, however, allows airborne sound to pass; hence there is an inherent problem here. Fortunately, in the case of masonry construction, there is seldom a problem in respect of sound transmission, since it is sufficiently massive.

PART 3

Selection of external walls

These aspects of the performance of external walls will be taken into account when selecting an appropriate option for the construction of dwellings. In reality, the viable, cost-effective options for low-rise dwellings are restricted to relatively few alternative forms. All will satisfy these requirements to a greater or lesser extent, and the selection is often based on familiarity with a given construction form. The relative lack of popularity of timber frame construction is partly the result of a lack of familiarity on the part of the property developer and on the part of the potential purchaser. As we move towards more sustainable, energy-efficient building forms this is likely to change, and the popularity of the timber frame dwelling may increase.

REVIEW TASKS

■ Explain what is meant by a building's SAP rating. Refer to the current UK Building Regulations and identify the components and features that impact upon the Target Emission Rate for new dwellings.

■ Explain what is happening in a cold bridging effect.

■ Undertake an Internet search for 'cavity wall insulation', and compare and contrast the different products that are available.

■ Undertake the same task for 'cavity closers'.

■ Visit the companion website at www.palgrave.com/engineering/riley1 to view sample outline answers to the review tasks.

7.2 | Traditional external wall construction

Introduction

■ After studying this section you should be familiar with the main forms of traditional external wall construction for low-rise housing.

■ You should be able to understand the reasons for the development of specific design features within each form and be able to relate these to generic performance requirements.

■ You should be fully aware of the technical details associated with timber frame and traditional masonry walls of both solid and cavity construction.

■ You should appreciate the relative advantages and disadvantages of each form and their differing features.

■ You should understand the implications for selection of external wall type upon other elements of the building such as roof, floors and foundations.

■ You should be able to critically appraise the various forms and, given specific scenarios, be able to select appropriate external wall solutions.

Overview

The external walls of dwellings provide for a number of specific performance requirements or functional needs. These have been explored in the previous section and
will not be reiterated here. However, it must be understood that the common forms
of external wall used in dwellings today have evolved specifically to fulfil these needs.
The forms of building that are utilised most often fall into two broad groupings:
'traditional' masonry walls and timber frame walls. These can be subdivided further,
and the specific features of each type will be examined in detail within this section.

Masonry walls

Masonry walls are formed of bonded stone, bricks or blocks of various materials.
Traditionally, the term 'masonry' related to working with stone, but is now taken to
include walls based on the assembly of modular units of a variety of materials. Most
commonly these include bricks and blocks made from clay or concrete, and less
commonly from calcium silicate. The assembly of these units is a highly skilled craft
and the degree of accuracy required to ensure structural stability results in very fine
levels of tolerance. Even with low-rise dwellings, which tend to have only one or two
storeys, it is essential that the verticality of the wall is maintained to avoid failure
by buckling. Figure 7.7 illustrates common bonding patterns.

The assembly of a masonry wall relies on the incorporation of several other
elements in addition to the bricks and blocks. These can be categorised as primary
elements and secondary elements. By this definition, primary elements are those
that are utilised to create the main structure, with secondary elements being those
items that are required to finish the structure.

Primary elements used to construct the carcass of the structure might typically
include such items as bricks, blocks, damp-proof courses, lintels and wall ties.
Secondary elements used in the finishing of the structure might include such
elements as air bricks, vents, copings and fittings of various kinds.

As previously noted, masonry wall construction is based on the assembly of a
large number of small components in the form of bricks or blocks. These are set and
secured in a bonding mixture known as mortar. This method of construction derives
from the historic need to assemble the units by hand, without the facility for the
use of mechanised plant. Hence the units used are of a size which is easily manipulated. It is worthwhile considering each of the main elements in a little more detail

Bricks

BS 3921 defines a brick as a 'walling unit not exceeding 337.5 mm in length,
225 mm in width and 112.5 in height'. The standard UK size is 215 mm ×
102.5 mm × 65 mm. This rather strange size is actually a result of the metric measurement of an imperial size of brick. Metric modular bricks are used elsewhere in
Europe, but their adoption in Britain has been slow. Bricks are formed from three
common materials, these being clay, concrete and calcium silicate.

Figure 7.7
Loadings applied to wall sections, and some common bonding patterns.

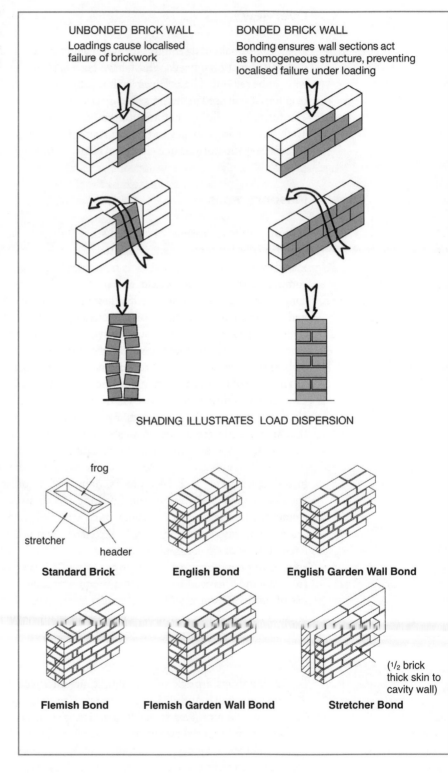

UNBONDED BRICK WALL
Loadings cause localised failure of brickwork

BONDED BRICK WALL
Bonding ensures wall sections act as homogeneous structure, preventing localised failure under loading

SHADING ILLUSTRATES LOAD DISPERSION

frog

stretcher

header

Standard Brick

English Bond

English Garden Wall Bond

Flemish Bond

Flemish Garden Wall Bond

Stretcher Bond

(½ brick thick skin to cavity wall)

Clay bricks

Clay bricks are manufactured from clay, shale or brickearth, formed to the requisite shape and fired in a kiln. The formation of the brick may take place by hand (although this is costly), by machine pressing or by machine wire cutting, each resulting in a different appearance. The classification of bricks is normally based on variety, quality and type. All of these aspects must be defined when specifying bricks. The various types will potentially differ greatly in their performance characteristics. Thus some bricks are more suitable for certain locations or uses than others.

Three varieties of clay brick are utilised in the construction process. The varieties are known as common bricks, facing bricks and engineering bricks.

Common bricks are suitable for general work, where aesthetics are not too important. Facing bricks are manufactured to give a high quality of surface appearance and to allow **fair faced finish** to unplastered walls. Engineering bricks are dense bricks with high compressive strength and low water absorption, allowing them to be used below ground or in areas of heavy loading. In older houses examples of all three varieties may be found, with facing bricks to the front elevations, common bricks to the rear elevations and engineering bricks at damp-proof course level. The use of engineering bricks as a damp-proof course is not permitted under modern Building Regulations, but in the Victorian era their use was common.

In addition to the varieties of brick there are three qualities of clay brick that are used. These are internal quality, suitable only for internal walls and partitions; ordinary quality, suitable for external work above ground level in conditions which are not subject to severe exposure; and special quality, suitable for conditions of extreme exposure or below ground.

The final characteristic for defining clay bricks is the type. Solid, perforated and cellular types of brick may be used, where:

- Solid bricks include holes through the brick that do not exceed 25 per cent of the brick volume. Frogs do not exceed 20 per cent of the brick volume.
- Perforated bricks include holes passing through the brick that exceed 25 per cent of the brick volume.
- Cellular bricks include holes that are closed at one end and exceed 25 per cent of the brick volume.

Calcium silicate bricks

Calcium silicate bricks are produced from a sand/flint base mixed with water and lime. They are often provided with colouring pigments to give variety. These bricks give a stark, modern appearance and are used in larger buildings more often than in housing, where a more traditional appearance is generally wanted. Unlike clay bricks, they are manufactured by moulding and autoclaving, giving very high degrees of accuracy and consistency in size and shape. Calcium silicate bricks are classified in three main categories. These are:

- 'Specials', which are suitable where great strength is necessary, or where they will be subjected to excessive moisture and/or freezing
- 'Class A', which are suitable for general external work
- 'Class B', which are suitable only for interior use.

Fair faced finish is the term applied to walls that are not intended to receive an application of plaster or render and are to be finished with the brick or block pointed to a good quality.

PART 3

Concrete bricks

Although they are now rarely used for house construction, calcium silicate bricks were popular in the 1970s. Their 'modern' appearance was considered fashionable for a short time. This is an example of the influence of people's taste upon the design of houses.

Such bricks are produced in the same manner as calcium silicate bricks, but using aggregates and ordinary Portland cement as the component materials. These are often used where high strength is needed without the requirement for the brick to provide an aesthetically pleasing finish.

Blocks

BS 6073 defines blocks as walling units larger than the sizes specified for bricks. In housing construction they are generally made from dense or aerated concrete. Although clay blocks are also available, they are rarely used in Britain.

The assembly of large numbers of independent units to form a continuous wall section presents difficulties in terms of loadbearing characteristics and structural stability (Figure 7.7). In order to overcome these difficulties they must be assembled in such a way as to achieve the performance of a single homogeneous unit. This is achieved by bonding of the bricks and blocks, together with ensuring that the **slenderness ratio** of the wall is within acceptable limits, as discussed earlier. The wall sections must be capable of withstanding lateral and vertical loadings. This is aided by the **bonding** of solid and cavity walls.

The nature of cavity wall construction is such that two slender leaves are formed next to each other, with an air gap between. This results in slender wall sections, which may become unstable. Hence, in order to ensure that they act as one broad unit, they must be tied together using wall ties. The ties must prevent the leaves from acting independently and ensure that the wall remains stable. The slenderness ratio of walls, that is, the proportion of the thickness of the wall relative to its height, is governed by the Building Regulations.

Figure 7.8 shows one of the earlier forms of cavity wall with both skins of brickwork. As bricks are now mainly produced to a standard height of 65 mm, the use of brick and brick cavity walls ensures that mortar joints will be at the same level on both skins and that incorporation of wall ties is easily achieved. However, more modern walls use blockwork inner skins, and these are of course significantly bigger (approximately 59 bricks per square metre compared with 10 blocks). A standard brick course is 75 mm high, comprising a 65 mm brick and a 10 mm mortar joint. By contrast, a block course is 225 mm high: a 215 mm block with a 10 mm mortar joint. One block course is therefore equivalent to three brick courses. BS 1243 suggests that in ordinary house cavity walls the standard spacing of wall ties is 900 mm horizontally and 450 mm vertically, with ties laid in a diagonal pattern. This means that ties will be built into every second block course or, if you prefer, every sixth brick course.

Clearly, bricks and blocks need to be of coordinating size if the mortar joints are to align to allow the building in of the wall ties.

The term **slenderness ratio** relates to the proportional dimensions between height and thickness of structural elements.

Bonding is the overlapping of bricks to ensure that load is dispersed. Several patterns are used, including 'English', 'Flemish' and 'Stretcher' bonds (see Figure 7.7).

Figure 7.8
Provision of wall ties in cavity walls.

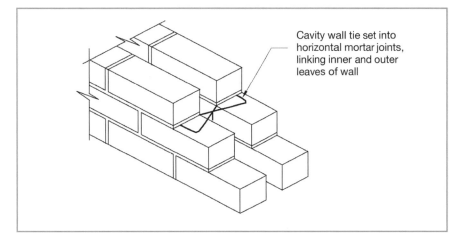

Cavity wall tie set into horizontal mortar joints, linking inner and outer leaves of wall

The nature of the environment in which wall ties are used is such that the outer leaf, which is almost permanently wet, provides a corrosive situation for the end of the tie. Wall tie failure is common in older buildings as a consequence; this is in the main due to ineffective anti-corrosion treatment. Modern ties must be formed of corrosion-resistant materials such as galvanised mild steel, stainless steel or, increasingly, plastics (Figure 7.9). The passing of a wall tie across the cavity may give rise to the passage of moisture. This is prevented by the incorporation into the design of the tie of an anti-capillary drip.

Figure 7.9
Forms of wall tie.

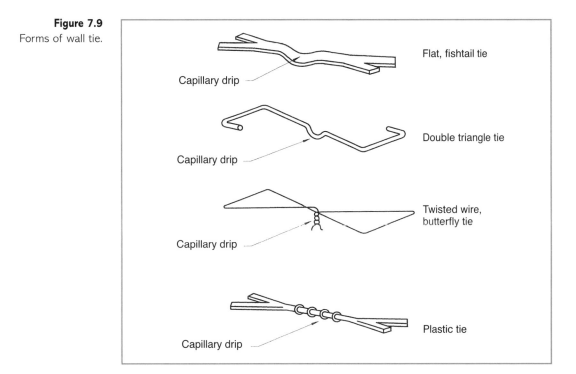

Flat, fishtail tie
Capillary drip

Double triangle tie
Capillary drip

Twisted wire, butterfly tie
Capillary drip

Plastic tie
Capillary drip

PART 3

Figure 7.10
Insulation of cavity walls.

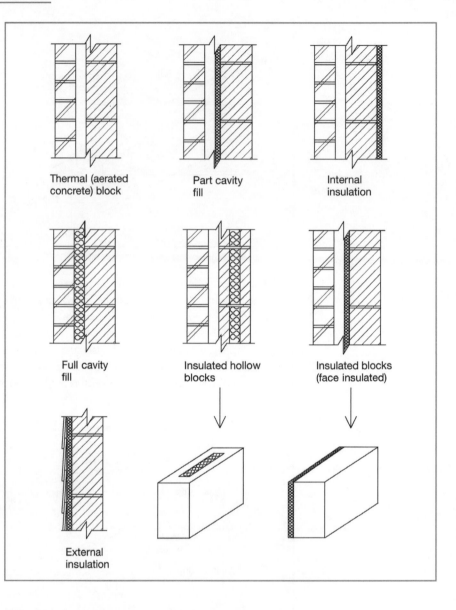

Thermal insulation

The evolution of the house external wall in more recent times (since the 1970s particularly) can be linked with the changes in thermal insulation standards required by the Building Regulations (Part L, Conservation of Fuel and Power). The U value is the typical way in which thermal insulation standards are set. The U value is also known as the thermal transmission coefficient and is measured in $W/m^2\,°C$ or using the absolute temperature scale $W/m^2\,K$.

Watts are the unit of radiant flux or flow of energy, and the Regulations attempt to limit the flow per square metre by setting maximum values for different external elements such as walls and roofs. The Building Regulations require a maximum U

Figure 7.11
Installation of insulation in
cavity walls.

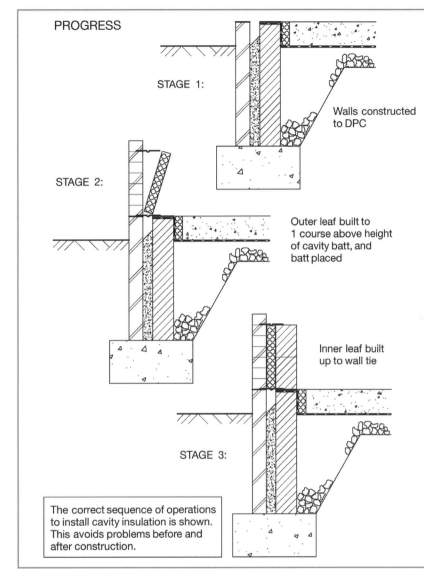

PROGRESS

STAGE 1:

Walls constructed
to DPC

STAGE 2:

Outer leaf built to
1 course above height
of cavity batt, and
batt placed

Inner leaf built
up to wall tie

STAGE 3:

The correct sequence of operations
to install cavity insulation is shown.
This avoids problems before and
after construction.

value of 0.30 W/m² K for walls and 0.20 W/m² K for roofs. This means that roofs have to be better insulated than walls. (Walls can lose almost half a watt per square metre of wall for each degree temperature difference between the inside and the outside of the building, while roofs would lose only one-quarter of a watt.)

When the U value of walls was improved some years ago to 0.60, this changed the way in which walls were put together, in that there was a significant pressure placed on the blockwork skin to provide good levels of insulation while remaining strong enough to carry the loads impressed on it. After the U value for walls changed to 0.45 it became very difficult to meet the thermal standards without incorporating insulation into the wall, and today partial cavity fill with rigid insulation boards (batts) is the norm. The drive for higher levels of thermal performance for walls has resulted in the need to develop even more effective methods of including insulation.

Figure 7.10 shows some of the various ways in which insulation may be incorporated into a cavity wall.

If cavity insulation is to be included in the wall it tends to be in the form of rigid boards fitted between the wall ties. Figure 7.11 shows this process where complete cavity fill is used. Where partial cavity fill is used, the cavity is maintained by securing the insulation against the inner leaf using clips attached to the wall ties.

Air permeability is also a major issue in the delivery of energy-efficient dwellings. As such, it is an important part of the delivery of an appropriate TER. In larger developments it is a requirement to demonstrate 'assessed air permeability'. This should not exceed 8.0 m^3 'assessed air permeability' per hour per m^2 (measured at 50 Pa). The detailing of the functions between walls and other elements along with the general quality of construction are important in minimising air permeability.

Movement accommodation

All building materials change their volumes to some extent when affected by changes in temperature and moisture, and masonry is no exception. If unrestrained, the effects of such movement can be considerable; hence movement accommodation joints must be incorporated to allow for such occurrences.

The coefficient of expansion of clay is surprisingly high, and occasionally this will mean the incorporation of movement joints in housing (Figure 7.12). Terraced rows of house units may be typically where the joints are required.

A heavy tie may be needed at the movement joint to reinforce this point, where the bond of the bricks is broken. As movement is expected at the joint, it is important to allow the wall to move while still taking support from the tie. Figure 7.13 shows some typical details.

Figure 7.12 Movement accommodation joints in brickwork (spacing).

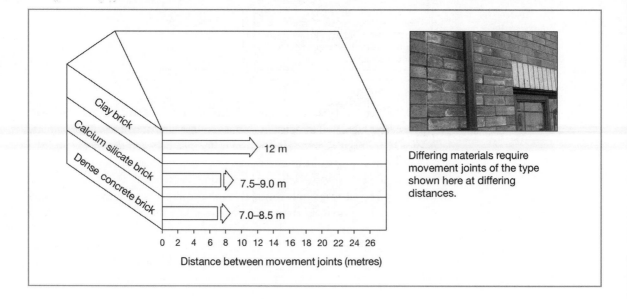

Clay brick — 12 m

Calcium silicate brick — 7.5–9.0 m

Dense concrete brick — 7.0–8.5 m

0 2 4 6 8 10 12 14 16 18 20 22 24 26
Distance between movement joints (metres)

Differing materials require movement joints of the type shown here at differing distances.

Figure 7.13
Movement joint with tie support.

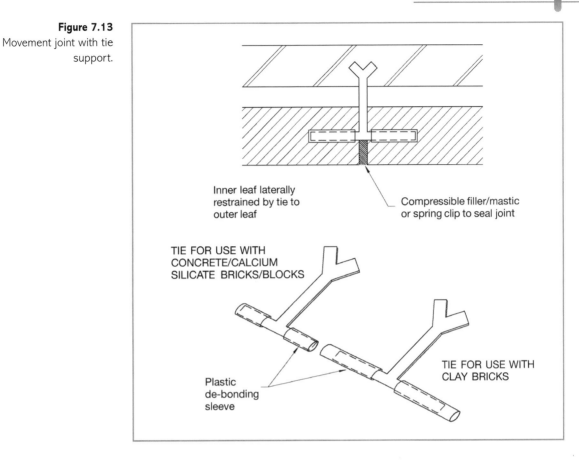

Inner leaf laterally restrained by tie to outer leaf

Compressible filler/mastic or spring clip to seal joint

TIE FOR USE WITH CONCRETE/CALCIUM SILICATE BRICKS/BLOCKS

TIE FOR USE WITH CLAY BRICKS

Plastic de-bonding sleeve

Figure 7.14
Options for pointing of brick and block walls.

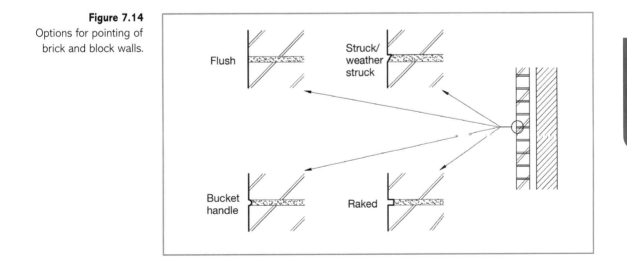

Flush

Struck/ weather struck

Bucket handle

Raked

PART 3

Surface appearance and pointing

The exposed face of the wall is likely to be finished 'fair faced'. The brickwork or blockwork is finished to a high standard and is not intended to receive any form of surface finish or decorative application. In order to achieve this, and to ensure that the surface is weatherproof, the mortar joints are finished by pointing. Several options are available and each will give a different appearance to the wall (Figure 7.14). In addition, they will perform differently in terms of weather resistance. Raked pointing gives a modern appearance, but performs less well than struck pointing in resisting the passage of moisture. Ultimately the selection is a matter of personal preference on the part of the designer.

REVIEW TASKS

- Why is dimensional coordination important when sizing bricks and blocks for use in an external wall?

- What is the standard height for a course of bricks and for a course of blocks?

- Consider the options available for thermal insulation and compare and contrast them in terms of relative advantages.

- Visit the companion website at www.palgrave.com/engineering/riley1 to view sample outline answers to the review tasks.

7.3 | Timber frame construction

Introduction

- After studying this section you should be able to refer to the principles of timber framed construction and how they differ from those for masonry construction.
- You should be familiar with the balloon and platform methods of timber frame.
- You should have gained an appreciation of the detailing which is necessary to protect the timbers from damage and which is needed to allow for the dimensional movement expected in the frame owing to the hygroscopic nature of timber.

Overview

In most traditional dwellings with masonry external walls, the wall is of cavity construction with an outer leaf of brickwork and an inner leaf of concrete block. In timber frame construction the concrete block inner leaf is replaced with a structural timber frame. This frame is the main loadbearing element of the external wall and the outer leaf of brick does not take any of the structural loads from the dwelling. The outer leaf of brickwork provides a weather shield and an aesthetically pleasing appearance. Indeed, the use of a brick outer leaf can be dispensed with, as a variety of other weatherproof claddings could be adopted. Examples of alternatives include timber or plastic weatherboarding, tile hanging and applied render finishes.

In recent years, timber frame houses have suffered from a poor public image. This arose as a result of failings associated with early examples. These failings were linked to poor detailing and site practice, resulting in moisture entering the timbers of the frame. In later examples, these problems have largely been eradicated and the use of timber frame construction now provides a viable, cost-effective method of construction for dwellings.

There are a number of distinct advantages associated with timber frame construction. Firstly, the nature of the form is essentially based on prefabrication; panels are formed in a factory environment and assembled as large units on-site (Figure 7.15). Hence there is the potential for high levels of quality control in manufacture, together with the benefit of economies of scale associated with 'mass' production. In addition, the flexible nature of the construction form lends itself to the manufacture of panels for bespoke designs. Indeed, there has been a great expansion in the level of use of timber frame for self-builders.

Secondly, the strength-to-weight ratio of timber is relatively high. Therefore the dead weight of a timber frame dwelling is significantly less than that of a traditional dwelling of comparable size. As a consequence there will be the potential for savings associated with lighter forms of foundation solution. In addition, the lower level of loading applied to the ground means that **strata** with lower bearing capacities may be utilised for construction of dwellings. This is a significant issue in terms of the **Developer's equation**, since less suitable sites will be cheaper to buy. It is also relevant to the issue of sustainability of the construction process, as there is increasing pressure to preserve greenfield sites.

Thirdly, because timber frame construction is largely a dry form of building, there is the benefit of accelerated progress on-site. Unlike traditional forms of construction, which require drying out time for the masonry walls and internal finishes, timber frame construction can be almost totally dry. Thus the timescale of construction can be reduced greatly.

Strata is the term used to describe the various layers that make up the ground on the site. The ground is generally made up of a number of layers of different soil types, varying with depth.

The **Developer's equation** was introduced in Chapter 2 (section 2.1). It is a simple relationship between land cost, income, outlay and profit.

Timber frame construction uses prefabricated panel sections to create the structural elements of the building. Here we see open wall panels stored on-site awaiting assembly.

Figure 7.15
Timber frame construction.

PART 3

'**Wet trades**' are those operations that traditionally involve the use of 'wet' materials such as cement, mortar, plaster and concrete. Increasingly, 'dry' alternatives are being used to remove the time loss required while these elements dry or cure, allowing subsequent operations to take place.

One advantage of timber frame is the fact that much of the work is dry compared with the '**wet trades**' of traditional construction. However, the major advantage is generally regarded as the speed of assembly resulting from prefabrication.

An aspect of timber frame construction is the need to carefully design for coordination of the elements and components – economies of scale can result from standardisation.

Principles of timber frame construction

The concept of timber frame construction is based on the erection of a loadbearing timber frame supporting the dead and live loads from upper floors, roofs and the timber frame wall itself. This structure is then clad with a weatherproof enclosure. In Britain the enclosure is most commonly formed in brick to mimic the appearance of a traditional dwelling. Two main forms of timber frame exist, termed platform frame and balloon frame. Although the balloon frame form is less common in Britain it is used extensively elsewhere. Both forms are based on the assembly of large prefabricated panels to form the structural external walls of the dwelling. These are then covered with a weatherproof cladding to provide an aesthetically pleasing appearance.

The construction of the individual panels is based on the provision of vertical studs, typically 50 mm × 100 mm, although sometimes larger. The studs are placed at 400 mm or 600 mm centres to coincide with the modular size of plasterboard sheets. The panel studs are fixed between top and bottom timber rails of the same dimensions. These are termed the 'head rail' and the 'bottom rail'. In addition, wherever it is anticipated that heavy fixtures will need to be supported, horizontal timbers, or noggins, will be inserted between the studs. It is therefore important that the positions of items such as wall cupboards and sanitary fittings are considered before assembly of the panels.

In order to provide a durable external face to the panel, and to ensure rigidity, a sheathing panel of plywood or bitumen-impregnated particle board, typically of 9.5 mm thickness, is nailed to the panel framing at 150 mm centres to ensure that the complete unit is braced against deformation. This is of particular importance in protecting the assembled panel units against the effects of wind loadings. In addition, the sheathing provides a continuous base for the attachment of cladding fixings such as battens for tile hanging or wall ties for external brick cladding.

The external face of the sheathing must be protected against the elements during the construction process. It is possible that the sheathing will be exposed for several weeks prior to the provision of the weatherproof cladding to the outside of the building. For this reason a protective covering is applied to the panels; this is termed a 'breather membrane' or a 'building paper'. The breather membrane fulfils several important functions. Firstly, it acts to protect the panels against rain during the exposed period of the construction process. Secondly, it acts to protect the panels against the possibility of wind-driven rain crossing the cavity. Thirdly, it is essential that any moisture that does find its way into the panels during construction or during the life of the property is allowed to escape. In order to facilitate this, it is important that the covering is vapour permeable, thus allowing any trapped moisture to escape as vapour to the outside – hence the term 'breather paper'.

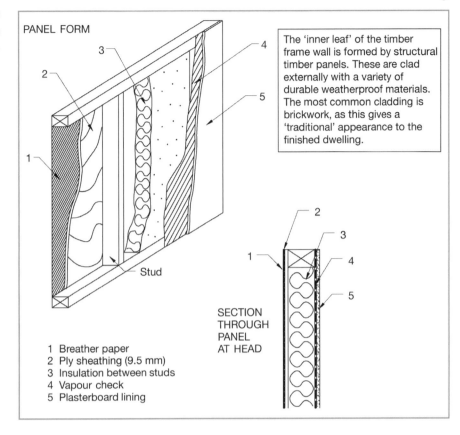

Figure 7.16
Typical ingredients of a timber framed wall panel.

PANEL FORM

The 'inner leaf' of the timber frame wall is formed by structural timber panels. These are clad externally with a variety of durable weatherproof materials. The most common cladding is brickwork, as this gives a 'traditional' appearance to the finished dwelling.

Stud

SECTION
THROUGH
PANEL
AT HEAD

1 Breather paper
2 Ply sheathing (9.5 mm)
3 Insulation between studs
4 Vapour check
5 Plasterboard lining

The void within the panels must be insulated to achieve the required level of resistance to the passage of heat. This is generally based on the installation of fibre quilt insulation material fitted tightly between the studs. The thickness of the panel dictates the level of insulation that is achievable. It is probable that in the future the panel thickness will increase to provide higher levels of thermal insulation. The installation of this insulation material results in a high level of temperature difference from the inside face to the outside face of the panel. This introduces the risk of interstitial condensation. Clearly this is a serious risk, as the structure is timber, and therefore potentially prone to decay. In order to minimise the risk of condensation within the panel it is important that a vapour check is provided at the inner face. This normally takes the form of polythene sheeting, which arrests the passage of moisture generated within the building into the core of the timber frame panel. It must be accepted that this is a vapour check rather than a vapour barrier. The reason for this is that the inner face of the wall will be pierced by fixings and holes to facilitate the installation of light switches, socket outlets and so on. Thus the vapour-resistant polythene is not continuous and will not resist the passage of moisture vapour totally. However, it will reduce it to a minimum.

The inner face of the panel is finished with the application of plasterboard sheeting, which may be dry finished or skimmed with a thin coat of plaster. The ingredients of a typical wall panel are shown in Figure 7.16.

Timber frame alternatives

Balloon frame

In the case of two-storey dwellings the balloon frame consists of wall panels that are the full height of the building (Figure 7.17). Following erection, the first floor joists are fixed to ledger timbers that are inserted between the studs of the panel. One disadvantage of this form of frame is the need to provide cavity barriers within the panels to arrest fire spread between floors owing to the continuous nature of the void within the panel. This contributes to the lack of popularity of this form. These cavity barriers are often formed from timber sections cut to size and fixed between studs on-site. The need to do this and to provide ledger timbers to support the upper floors reduces the benefit of prefabrication and increases the labour input on-site. Hence the construction process is slowed down and the potential for on-site inaccuracies is increased. The main advantage of this frame type is that the external walls are erected independently of the upper floors. Therefore the building envelope can be erected to full height and the roof formed in a very short time. In this way a weatherproof enclosure is created quickly, allowing the work internally to be undertaken in a protected environment.

A feature of the balloon frame is the two-storey high wall panels, which are shown in Figure 7.18.

Platform frame

The most common type of timber frame construction adopted for dwellings in Britain is the platform frame (Figure 7.19). Unlike the balloon frame this relies on the use of panels that are the height of a single storey. This results in the creation of panels that are very easy to handle during transit and assembly. The ground floor panels are erected first, with individual panels linked together using a timber head binder. After this, the upper floor is formed, including the deck, resting on the head binder. At each end of the floor joists, where they are supported on the frame panel, the floor void is closed off using a header joist. The completed lower section of the building then acts as a platform for the formation of the upper storey, which is formed in the same way.

The platform frame has a number of advantages. Firstly, the panels are easily transported and handled on-site because of their convenient size. Secondly, the provision of the upper floor at a relatively early stage allows its use as a working platform for the subsequent operations. Thus the need for temporary working platforms is reduced. However, the formation of the floor at this early stage also results in a disadvantage in that it interrupts the operation of forming the external structural walls. Hence the roof is formed later and the building cannot be made weathertight as soon, and there is the potential for a longer period of exposure to the elements. Notwithstanding this, platform construction accounts for the bulk of timber frame construction in Britain.

It can be clearly seen that use of the platform frame allows the upper floor to sit directly on top of the lower wall panel (Figure 7.20).

Figure 7.17
Balloon frame
construction.

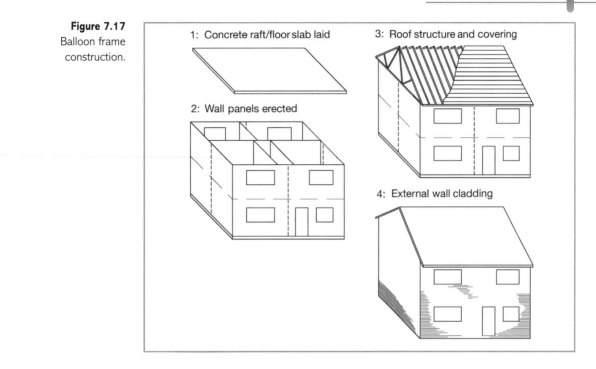

1: Concrete raft/floor slab laid

2: Wall panels erected

3: Roof structure and covering

4: External wall cladding

Figure 7.18
The two-storey high wall
panels of the balloon
frame.

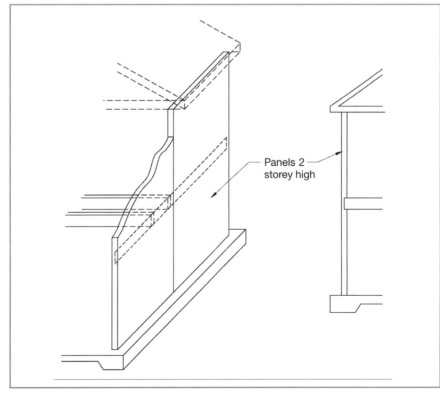

Panels 2
storey high

Figure 7.19
Platform frame
construction.

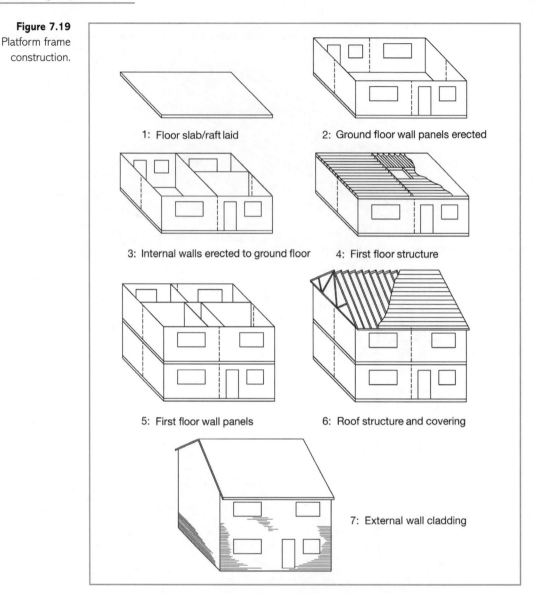

1: Floor slab/raft laid

2: Ground floor wall panels erected

3: Internal walls erected to ground floor

4: First floor structure

5: First floor wall panels

6: Roof structure and covering

7: External wall cladding

Other variants of the timber frame form exist, although their use is not widespread in Britain and they broadly share the same features as those described here. See also Figure 7.21.

REVIEW TASKS

■ List *two* features of timber framed construction which are included to prevent moisture entry to the timber.

■ What distinguishes the balloon frame technique from the platform technique?

■ List the benefits of timber framed construction in terms of sustainability.

■ Visit the companion website at www.palgrave.com/engineering/riley1 to view sample outline answers to the review tasks.

Figure 7.20
The single-storey wall panels of the platform frame.

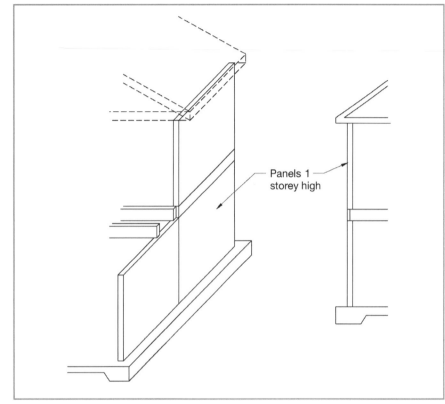

Panels 1 storey high

Figure 7.21
Use of scaffolding in erection of timber frame walls.

Unlike traditional masonry walls timber frame walls do not possess sufficient stability during erection to allow scaffolding to be erected as work proceeds. Instead the scaffolding is erected first and is used as a mechanism to provide temporary bracing to the frame during erection.

Assembly of timber frame buildings

As stated previously, the most common form of timber frame construction in Britain is the platform frame. Although there is some degree of variation in the assembly of the various forms of framing, they are essentially similar. Hence we shall consider only the platform frame as an illustration of the broad principles involved. The sequence of operations involved in the assembly of a timber frame dwelling has already been outlined. It is now worthwhile focusing on some of the important assembly details involved. The main areas of importance are the junctions between panels and the connections between wall panels and the foundation/ground floor, the upper floor and the wall/roof junction.

Junctions between panels

In some instances the full length of the external wall may be formed by a single panel. This is unusual, and it is more common for walls to be made up of a number of individual panels linked together (Figure 7.22). The panels are generally butted together and nailed with the addition of a head binder running across the top of the panels to ensure alignment. The head binder also creates a double rail at the top of the ground floor panels, providing a sound base for the formation of the upper floor. At the base of the wall the bottom rail is fixed to a continuous timber sill plate or sole plate. Where panels are joined in continuous lengths the connections are simple in nature. At corners, where adjoining panels are perpendicular, it is necessary to introduce additional studs to facilitate connections. An alternative is to manufacture panels specifically for these positions. This reduces the level of standardisation, and hence the degree of cost-effectiveness. Consequently the addition of extra studs on-site is far more common.

Connections to foundations

Timber frame construction is lighter in form than traditional masonry construction. Hence the design of the foundations takes this into account. However, the foundation forms are essentially the same as those available for traditionally constructed houses. In practice there are two commonly used foundation solutions: the strip foundation and the reinforced raft foundation. Where the strip foundation is used it is most common to adopt a ground-supported concrete floor slab also. In both of these scenarios the method of connecting the timber wall panels is the same.

A timber sill or sole plate, which has been vacuum impregnated with preservative, is secured to the lower wall section or the raft. This is set onto a length of damp-proof course and is bedded on cement mortar to provide a firm, level base for attaching the wall panels. In the past the sole plate would have been secured to the wall or raft by shot fixing or by drilling holes through the timber and inserting anchor bolts. In both of these cases the untreated core of the timber is exposed and the DPC is pierced. Thus there is the potential for moisture to pass into the timber, possibly resulting in decay. In addition, where anchor bolts are used it is necessary to either countersink the bolt heads or recess the bottom rail of the wall panel to avoid fouling.

Figure 7.22
Panel junctions in timber
frame construction.

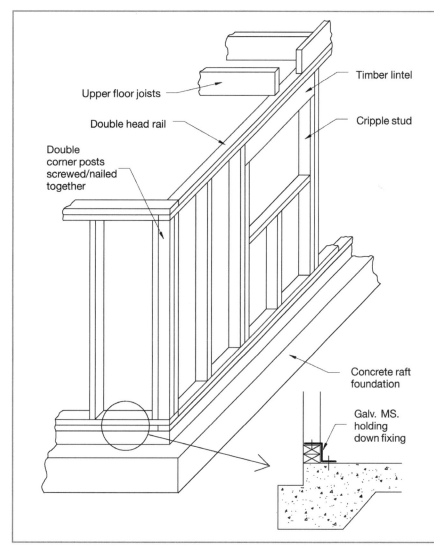

More modern techniques adopt the use of a non-ferrous holding-down strap that does not pierce the DPC or pass into the core of the timber. Figure 7.23 illustrates two possible options.

Having secured the sole plate, the wall panels are nailed directly to it.

Connections to upper floors

The upper floor construction used in timber frame housing is the same as that used in traditional construction. The mode of connection to the wall panels is rather specific, however. The floor joists bear directly onto the head binder of the lower wall panels (Figure 7.24). The use of the head binder together with the head rail of the panel results in a robust double section providing the support for the joists (Figure 7.25). It is common, though not essential, for the joists to line up with the

Figure 7.23
Connections to
foundations.

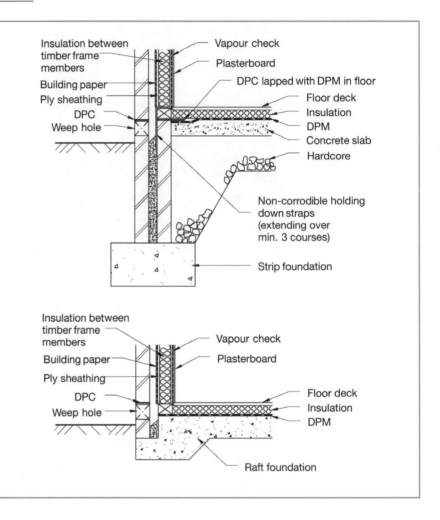

Insulation between timber frame members
Building paper
Ply sheathing
DPC
Weep hole

Vapour check
Plasterboard
DPC lapped with DPM in floor
Floor deck
Insulation
DPM
Concrete slab
Hardcore

Non-corrodible holding down straps (extending over min. 3 courses)

Strip foundation

Insulation between timber frame members
Building paper
Ply sheathing
DPC
Weep hole

Vapour check
Plasterboard
Floor deck
Insulation
DPM

Raft foundation

Figure 7.24
Timber frame housing –
upper floor construction.

Here we see the connection between the upper floor of the dwelling and the external wall panels. Note that the gaps between the joists are filled with timber sections to prevent the passage of fire. Note also the double head rail to support the joists

Figure 7.25
Timber frame junction
details.

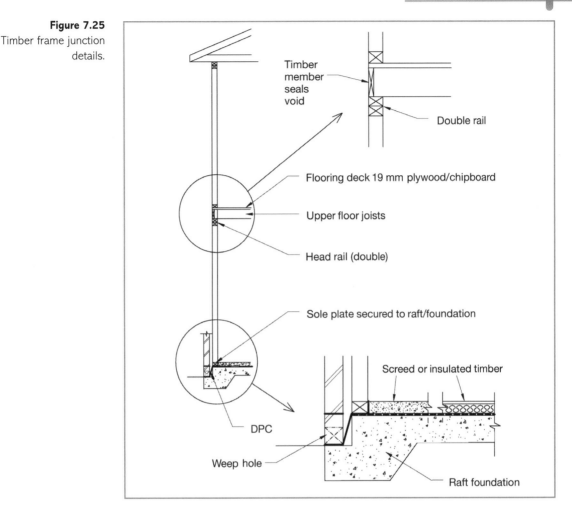

studs of the wall panels. The spacing of both studs and joists is generally either 600 mm or 400 mm; this maximises the efficiency of using modular plasterboard sheets for wall and ceiling finishes. After fixing the floor joists a header joist is provided around the edge of the floor to close the floor void and resist the passage of fire. At this stage the floor decking is laid up to the outside edges of the floor/wall assembly. The completed floor provides the platform for the next lift of wall panels, which are connected by nailing through the bottom rail into the floor.

Connections to roofs

The connection at roof level (Figure 7.26) is similar to that at the upper floor. A head binder is provided to connect the head rails of the individual wall panels and to provide a double rail base for supporting the roof. In timber frame construction this is almost certainly a trussed rafter roof structure. The individual trusses are connected to the head binder using galvanised steel saddle connectors. This is effected in exactly the same manner as the connection to the wall plate used in traditional construction.

PART 3

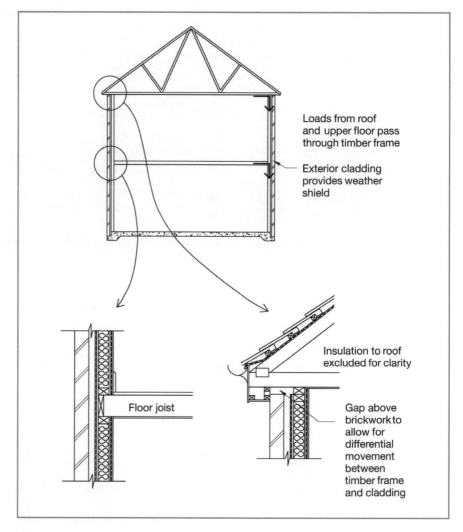

Figure 7.26
Floor and roof
connections.

Loads from roof
and upper floor pass
through timber frame

Exterior cladding
provides weather
shield

Floor joist

Insulation to roof
excluded for clarity

Gap above
brickwork to
allow for
differential
movement
between
timber frame
and cladding

Claddings to timber frames

A wide variety of exterior cladding options exist for timber frame buildings. When dealing with dwellings there is a tendency to attempt to mimic the more traditional masonry construction form. Hence the most common cladding option tends to be masonry, and in particular brick. Alternatives to this include a range of lightweight claddings such as horizontal or vertical boarding, tile hanging or rendered finishes, although these are far less popular. In all cases it is important that the cladding is spaced away from the timber wall panels to allow for venting and draining of the space. It is also essential to take into account the potential for differential movement between the structural frame panels and the selected cladding. The general principles involved are the same for all claddings, but the specific details may vary considerably.

Figure 7.27
Masonry cladding to
timber frame
construction.

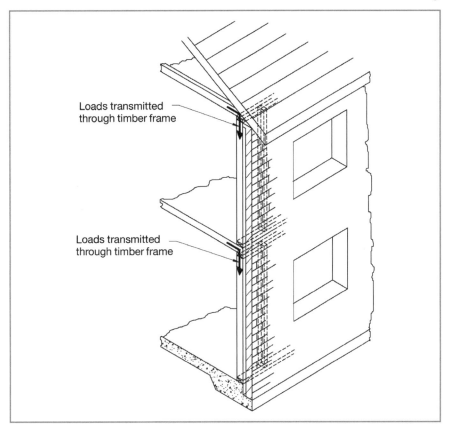

Loads transmitted
through timber frame

Loads transmitted
through timber frame

Masonry cladding

When masonry is used as the exterior cladding to timber frame buildings (Figure 7.27) it acts to a large extent in the same way as the outer leaf of a traditional cavity wall. It protects the inner frame from the elements and prevents moisture from entering the building by the creation of a capillary break, in the form of a drained cavity. The cavity should be at least 40 mm wide. In timber frame construction the exterior cladding does not take any loads from the building structure other than its own self-weight. The stability of the slender masonry leaf must be maintained by introducing wall ties that restrain the outer leaf by tying it to the timber frame. A common defect in older examples is the positioning of these ties such that they are nailed to the sheathing of the panel rather than to the studs. This does not allow for adequate restraint, and it is now accepted that the ties must be fixed through the sheathing into the studs of the wall panels. An added complexity is introduced as a consequence of the potential for differential movement between the timber frame and the masonry cladding. It is quite possible that the timber structure will be affected by a degree of shrinkage in the period shortly after construction. In addition, the rates at which the different materials will expand and contract when affected by changes in air temperature and humidity levels vary. The ties used to restrain the masonry must allow for this differential movement and must, therefore, have a

PART 3

degree of flexibility. Flexible stainless steel ties are the best solution. These should be fixed to the sheathing to coincide with the positions of studs at centres not greater than 600 mm horizontally and 450 mm vertically. As in the case of traditional cavity walls, they should fall slightly to the outside leaf.

Lightweight claddings

A range of lightweight claddings may be used to provide the weather shield to timber frame buildings. These are generally fixed directly to the frame panels using battens or spacers to create a slim cavity for draining and venting. The cavity should be at least 19 mm wide and should as far as possible be uninterrupted vertically.

Possible alternative options for cladding to timber framed housing include tiling, panel systems and rendered finishes, among others. The details of masonry, vertical tiling and possible alternative design solutions are illustrated in Figure 7.28.

Figure 7.28
Timber frame cladding
options

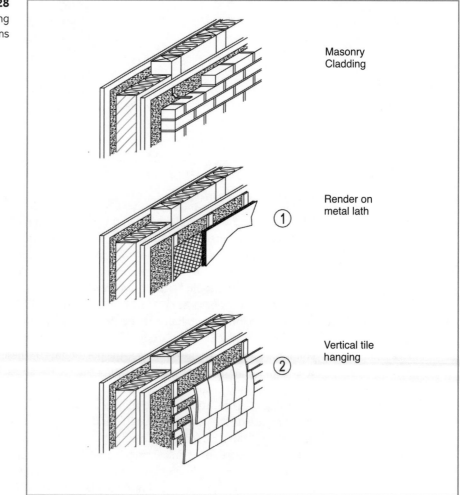

Timber frame and fire

In order to control the spread of smoke and flame through the concealed voids and cavities within timber frame buildings it is necessary to install **cavity barriers**. These are introduced at key locations within the building and must be capable of providing resistance to fire of at least 30 minutes. There are several methods of forming these barriers, but the most common are timber battens or preformed tubes of mineral wool insulation material. In addition to these cavity barriers, fire stops are required where there are a number of dwelling units within a single structural block. Hence in semi-detached and terraced housing fire stops are provided at the junctions of roofs and front and rear elevations with the party walls (Figure 7.29). The provision of these fire stops ensures that each individual unit forms a separate compartment capable of containing a fire for a period of 60 minutes. The typical location of cavity fire stops is shown in Figure 7.30.

Figure 7.29
Cavity fire stops.

The provision of fire stops in the cavity is evident here. The stops are fixed to the outer face of the wall panels prior to cladding with the brickwork outer leaf.

Figure 7.30
Fire stops.

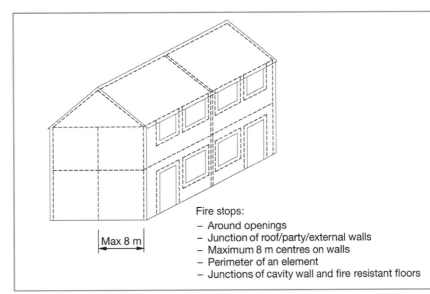

Max 8 m

Fire stops:
- Around openings
- Junction of roof/party/external walls
- Maximum 8 m centres on walls
- Perimeter of an element
- Junctions of cavity wall and fire resistant floors

PART 3

Figure 7.31
Accommodation of
movement – eaves
provision.

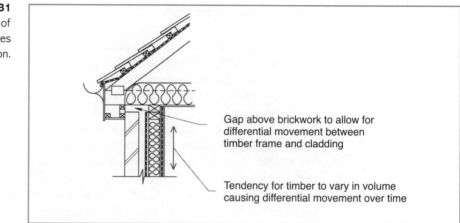

Gap above brickwork to allow for
differential movement between
timber frame and cladding

Tendency for timber to vary in volume
causing differential movement over time

In order to achieve this level of fire resistance it is usual for the party wall between dwelling units to be formed as two separate leaves with a gap of 50 mm between. Each side of the wall is then clad with two or three layers of gypsum plasterboard and mineral wool is fixed at the junctions with adjoining elements of structure. Increasing the mass of the party wall in this way also assists in providing an adequate level of resistance to the passage of sound.

Movement accommodation

Previously within this section the issue of differential movement between the timber frame and masonry cladding was noted. This is accommodated by the use of flexible ties as previously described, but it is also important that a series of other features is incorporated into the building to cope with this movement. It is probable that in the early part of the life of the building the timber frame will suffer slight shrinkage. If allowance for this is not made, the external leaf will become subject to the application of loads at contact points. The particular areas of concern are the eaves of the roof (Figure 7.31), around window openings and at junctions of different cladding materials. The risk of applied loads to the cladding is controlled by providing gaps to accommodate the potential movement.

REVIEW TASKS

■ Identify *two* features that are incorporated to accommodate frame movement in timber framed house designs.

■ What level of fire resistance is required by a dwelling? What features are used in timber frame construction to ensure that the fire regulations are satisfied?

■ Examine the planning applications in your local areas – this can be done via the Internet. How many timber frame house developments can you identify?

■ Visit the companion website at www.palgrave.com/engineering/riley1 to view sample outline answers to the review tasks.

CASE STUDY:

TIMBER FRAMED CONSTRUCTION

Traditional mass concrete strip foundations are formed to support the external walls. It is also common for raft foundations to be utilised for timber framed house construction. This site has several areas of sloping ground, and stepped foundations are utilised to accommodate this.

Timber framed panels are placed in position on the foundations/floorslab. Note the DPC material beneath the sole plates of the timber framed units, which is essential to prevent moisture penetration.

TIMBER FRAMED CONSTRUCTION (*continued*)

The timber framed units sit in the same position as the inner leaf of a traditional cavity wall with the cladding formed using an external skin of brickwork laid on a DPC near ground level. Note the reflective, insulative sheeting that is fixed to the outer face of the panels.

The upper floor structure is formed using timber and steel open-web joists spanning between the structural panels. Internal partitions are formed using timber stud panels. Note the multiple head rails on the structural wall panels to provide suitable bearing for the open-web joists.

Upper floors are covered with oriented strand board and provide an interim working platform for construction of the upper sections of the dwelling. Platform construction is used with single-storey height panels seen here stored at first-floor level, awaiting assembly.

The upper sections of the walls are formed, including profiled panels for the gable sections. Note the red breather paper covering to the external face of the gable sections. The panels are covered with reflective, insulative sheeting prior to cladding.

TIMBER FRAMED CONSTRUCTION (*continued*)

The roof structure is formed using trussed rafters. Note the temporary diagonal and longitudinal bracing used during construction, fixed to the outer edges of the trusses. Note also the permanent bracing that is secured within the roof structure to ensure stability of the roof and resistance to lateral loads.

Here we see the details at the gable of the roof where a 'gable ladder' is used to project beyond the wall and to provide a fixing for the barge boards that will be fitted later. The party wall between adjacent dwellings is also shown. Note the two adjacent wall panels, secured to each other with metal straps, whilst maintaining physical separation to prevent passage of noise. Effective fire-stopping will also be applied at this location.

The dwellings are nearing completion with brick cladding and traditional tiled roofs providing an effective weathershield for the properties. Once complete, the appearance is the same as a traditionally built house.

7.4 | Light steel frame construction

Introduction

- After studying this section you should appreciate the principles of house construction using light gauge steel frames.
- You should understand how this form of construction differs from timber frame and masonry construction.
- You should be familiar with the main methods of steel frame construction.
- You should have gained an appreciation of the detailing of this form of construction.

Overview

The broad principles of house construction using light gauge steel frame techniques are similar to those adopted for timber frame construction.

Principles of steel frame construction

The form of light steel framing that is utilised in the construction of houses is generally based on the standard C- and Z-sections that are manufactured by the cold-rolling of light gauge steel. These are quite unlike the heavier steel sections used for commer-

Figure 7.32
Stick-build frame.

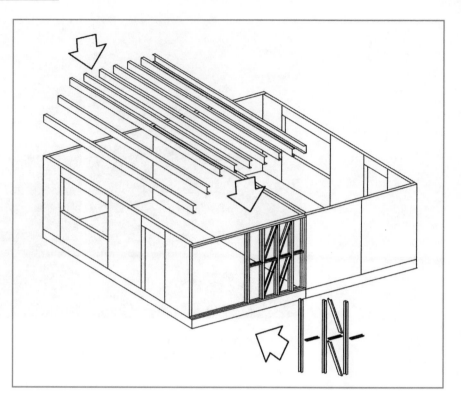

Figure 7.33
Assembling stick-build or panel frames.

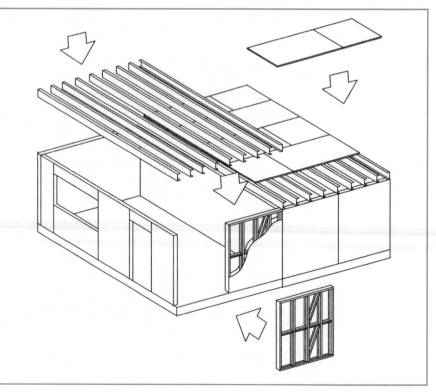

cial and industrial buildings, which typically use hot rolled I-section beams and columns. The light steel sections are galvanised to ensure durability and are assembled to form the structural frame of the building. Several options exist for the construction form: the steel members may be assembled in a factory environment to form panels which are fixed together on-site; alternatively, they may be assembled on-site using a variety of methods and design concepts. The main options available are:

- Stick-build frames
- Panelised frames
- Modular (volumetric) construction.

Stick-build frames

This form of construction relies on the assembly of individual sections to form struts, columns, beams and bracing that combine to create the structure to which the internal lining and external cladding are fixed (Figure 7.32). It is generally the case that the steel sections are delivered to site pre-cut to prescribed lengths, with accommodation holes for services and so on already introduced, and with the individual sections galvanised after cutting. The assembly of the individual components then relies on site-fixing using self-drilling, self-tapping screws or using nuts and bolts (Figure 7.33).

This is quite a labour-intensive process, which would seem to conflict with the concept of prefabrication and fast assembly. However, the process does lend itself to individual or one-off buildings which may be complex in form and which do not lend themselves to prefabrication.

The advantages of stick-build construction include:

- Facility for individual buildings and complex designs
- Simplicity of connection and construction
- Avoidance of factory set-up for low volume construction
- Ease of handling of components on-site
- Easy accommodation of variations and site tolerances
- Ease of transportation of densely packed members.

Panelised frames

The use of panelised frames shares much in terms of overall concept with the established forms of timber frame house construction. Wall panel sections are manufactured off-site, which may be delivered to site as open (Figure 7.34) or closed panels. Open panels comprise the structural skeleton of the frame to which internal elements such as insulation and external cladding and linings are added on-site. Closed panels comprise the structural skeleton and the internal components and external cladding and lining, all pre-assembled in the factory environment. The panels are connected on-site using the same techniques that are adopted for stick-build, with the use of bolts or self-drilling, self-tapping screws. Because of the high levels of accuracy that can be achieved using jigs in a factory environment, the panels can be manufactured with high quality levels. This aids rapid assembly on-site,

PART 3

Figure 7.34
Open wall panel sections.

The assembled open wall panel sections are seen here, loaded into carrying racks for transportation to the construction site.

reducing construction times significantly at the same time as increasing overall quality assurance.

The fine tolerances that result do, however, demand high levels of accuracy and consistency in the formation of foundations if the potential benefits in terms of speed and quality are to be achieved. As with timber frame construction, panels may be formed using either balloon or platform formations (Figure 7.35).

The advantages of panelised construction include:

- Enhanced quality control and reliability of assembly
- Reduced skills requirements on-site
- Reduced labour costs
- Economies of scale for large volume production of repetitive sections.

Modular (volumetric) construction

The principles of modular or volumetric construction are covered in Chapter 3. However, the use of steel frame for the creation of modular or volumetric buildings is also worthy of note separately here. The creation of buildings using modular form relies on the manufacture or assembly of three-dimensional modules, often based on whole sections of the building (Figure 7.36), or individual rooms or 'pods' that may be installed within the building enclosure (Figure 7.37). The units are manufactured at the factory and may be finished to varying degrees of completeness, from simple box forms to completely finished and decorated modules, complete with plumbing, electricity and associated fixtures and fittings. The units are stacked on-site to form the final structure of the building. This is generally provided with a cladding envelope, which may be a traditional-looking brick skin, to provide aesthetics and weatherproofing. This form performs best where there are high levels of repetition in the building form or where individual buildings are formed from repeated, standard modules. Hence the adoption of this technique for flats, student

Figure 7.35
Platform and balloon
construction of panels.

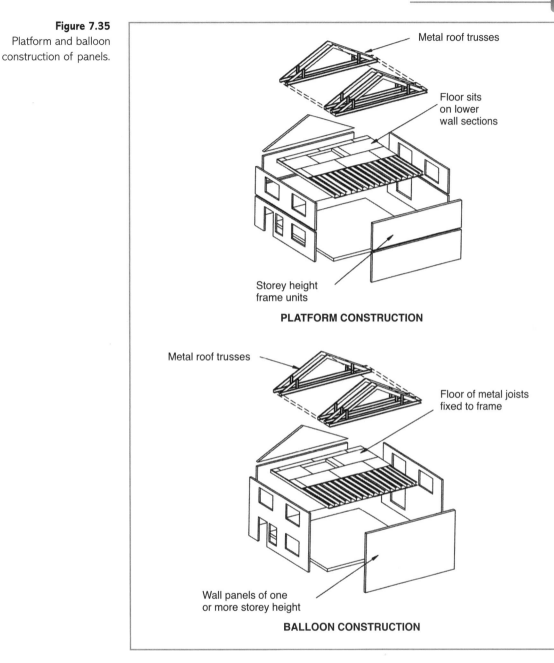

Metal roof trusses

Floor sits
on lower
wall sections

Storey height
frame units

PLATFORM CONSTRUCTION

Metal roof trusses

Floor of metal joists
fixed to frame

Wall panels of one
or more storey height

BALLOON CONSTRUCTION

accommodation and hotels is becoming more widespread.

Advantages of modular construction include:

- Quality assurance due to factory manufacture
- Greatly reduced construction time on-site
- Reduced skills requirements on-site
- Accelerated weather enclosure.

Figure 7.36
Volumetric construction.

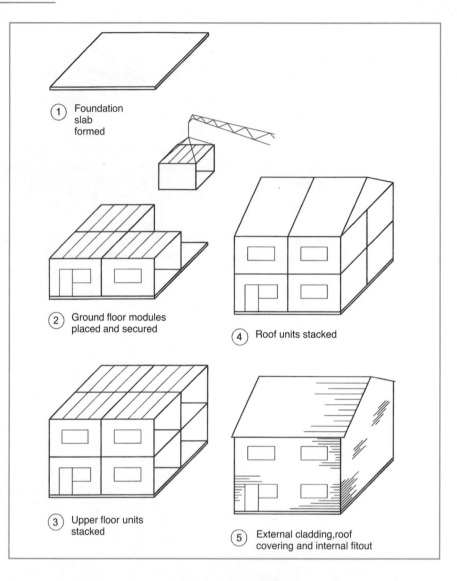

1. Foundation slab formed

2. Ground floor modules placed and secured

3. Upper floor units stacked

4. Roof units stacked

5. External cladding,roof covering and internal fitout

REVIEW TASKS

■ Consider the differences between stick-build, panelised and modular construction of steel frames.

■ Prepare a list of criteria that you would use to compare and contrast the three methods.

■ Apply these criteria to a project that you are familiar with and generate a comparison between the alternative approaches, indicating the benefits and disadvantages of each.

■ Visit the companion website at www.palgrave.com/engineering/riley1 to view sample outline answers to the review tasks.

Figure 7.37
Installation of individual
rooms within the
structure.

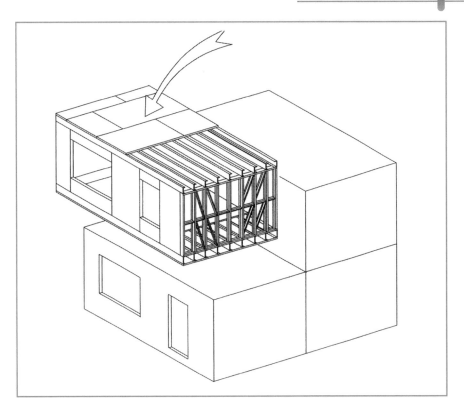

Assembly of steel frame buildings

The most common form of steel frame in the UK is panelised construction. There is a degree of similarity between the various systems that can be adopted, and the broad principles of assembly are to some extent generic. In this section the assembly principles of panelised construction will be referred to as an exemplar of the generic steel frame form. Where appropriate, or where there are significant differences, reference will also be made to other options. The general procedure for the erection of a steel framed house is illustrated in Figure 7.38. The procedure can be summarised as follows:

1. Formation of foundations/floor slab (or raft)
2. Placement of ground floor frames/panels
3. Placement of floor joists/cassettes and formation of openings
4. Placement of upper floor frames/panels and placement of 'wind girders' (bracing)
5. Placement of roof trusses
6. Cladding and finishing of building envelope.

Foundations

The foundations for steel framed housing are broadly the same as those used for traditional construction or timber frame construction. Accuracy of setting out and

PART 3

Figure 7.38
Erection procedure for
light steel frames.

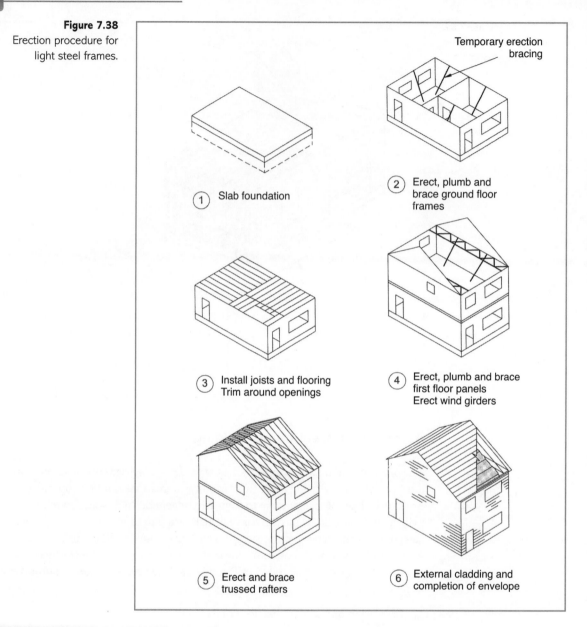

Temporary erection bracing

1 Slab foundation

2 Erect, plumb and brace ground floor frames

3 Install joists and flooring Trim around openings

4 Erect, plumb and brace first floor panels Erect wind girders

5 Erect and brace trussed rafters

6 External cladding and completion of envelope

formation is one of the key requirements for foundations of steel framed buildings, which are manufactured with high levels of accuracy. The ability of the frame contractor to place the frame quickly and securely relies on the foundations being formed within relatively fine dimensional tolerances.

As with all building types, the foundations will be designed to suit the local ground conditions, ensuring that the safe bearing pressure is not exceeded. The loads from steel framed houses are lower than for the equivalent building using traditional construction techniques. The use of strip and trench-fill foundations is common, although raft foundations, piles and ground beams and precast foundation systems may also be used.

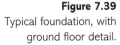

Figure 7.39
Typical foundation, with
ground floor detail.

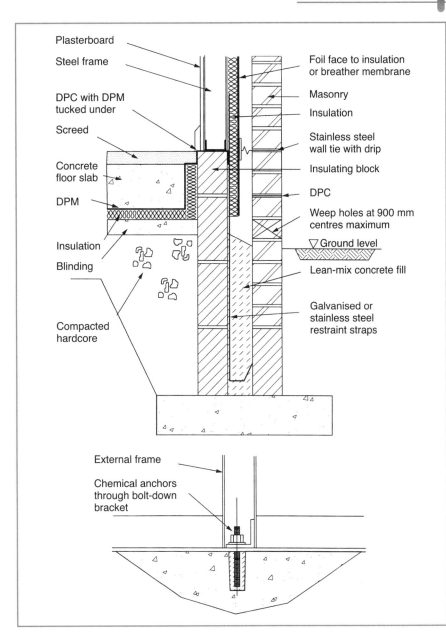

The interface between the frame and the foundation is one of the key areas that ensures durability and structural stability of the completed building. Although the steel frame is galvanised for protection against corrosion, to ensure durability it must still be subject to separation, as far as possible, from moisture rising from the ground. This is achieved by the placement of a robust damp-proof course beneath the wall panel, securely lapped with the damp-proof membrane under the ground floor. Figure 7.39 illustrates a typical foundation and ground floor detail. One of the important features of the detail is the use of restraint straps (or holding-down straps)

to secure the frame to the walls below ground and/or the foundation. The purpose of these straps is to resist the uplift generated by wind action upon the frame. Where raft foundations are utilised, these straps are replaced by holding-down bolts embedded in the concrete raft/ground floor slab.

Ground floors

The potential ground floor solutions for steel framed housing resemble those used for traditional construction. The most common form is the ground-supported concrete slab, which may be reinforced, depending on ground conditions. If a raft foundation is utilised, the floor and foundation are integral within the raft. The details that are commonly used to exclude moisture from the building in traditional construction, such as the provision of a DPM linked to the DPC in the external walls, are also found in steel frame construction. Similarly, the features used to ensure thermal insulation of the floor are the same as those adopted in traditional construction.

Walls

The walls to steel framed houses can be classified as loadbearing or non-loadbearing. The lightweight structural form of the panels that make up most of the walls is essentially similar to a stud partition. Those walls that are designed to take significant axial loads from the upper floors and roof are defined as loadbearing or 'structural' walls (Figure 7.40). Those that are intended to provide division of space and support their own self-weight would be considered as non-loadbearing. Whether the steel framed construction is based on stick-build, panelised or volumetric design, the composition of the wall panels will be broadly the same. The differences in these three approaches relate to the extent of prefabrication off-site rather than to the general assembly of the components that make up the walls.

Like timber frame construction, the steel frame is most commonly used to provide a structural frame for the building, which is then clad externally to ensure the provi-

Figure 7.40
Connection of external wall to internal partition and floor.

Here we see the connection of the external wall sections to an internal partition and the ground floor slab. Note the diagonal bracing, formed in flat steel sections, in contrast to the C-profile of the sections that make up the wall panels.

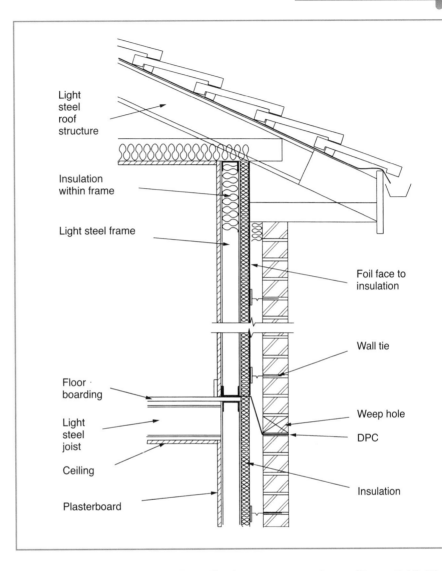

Figure 7.41
Steel structural frame for a building.

sion of a weatherproof and aesthetically pleasing outer enclosure (Figure 7.41). The construction provides a framed cavity wall; the steel frame forms a structural inner leaf, often with a traditional brick outer leaf providing weather-proofing. The wall panels are assembled using C-section galvanised steel studs, rails and noggins to create a skeleton structure which is clad internally and externally. The studs will be placed at centres of between 400 mm and 600 mm, depending on loading and configuration. The internal lining will be formed using foil backed plasterboard; this ensures that there is a vapour check between the board and the steel studs. The outer face of the panel may be provided with boarding or insulation with a breather membrane on the outer face. The void within the panel structure may also be filled with insulation material. In instances where the cavity between the steel frame and the outer cladding is provided with insulation, the wall is referred to as a 'warm frame' construction (so called because the frame is on the warm side of the insulation).

Figure 7.42
Bracing to steel frames.

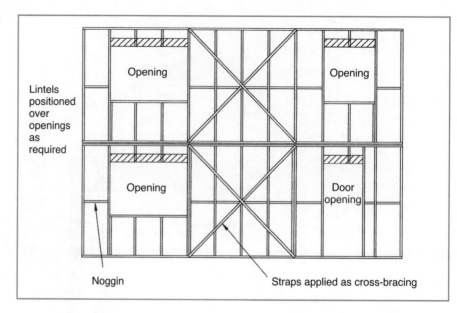

Openings for windows and doors will require the incorporation of lintels within the panels or frame sections (Figure 7.42). These will generally take the form of C-shaped lintels close to upper floor level, where there is the facility to tie to the structural floor to assist with lateral restraint. These will be supplemented by separate lintels for the support of the outer skin of the wall. In order to cope with the loads that may arise from wind action and in order to maintain the structural integrity of the panels, it is also necessary to incorporate bracing in the form of diagonal herringbone or cross-bracing.

Upper floors

Upper or intermediate floors will be of suspended form and may utilise cold-rolled steel sections or, for longer spans, lattice trusses or girders. Since the strength-to-weight ratio of steel floor joists is greater than that of timber of equivalent size, the floor can be a relatively lightweight assembly. In modular construction this may take the form of a 'cassette, which incorporates the upper surface of the floor and the ceiling finish beneath as a complete assembly. Holes for the passage of services will be pre-formed in the joists. If lattice sections are utilised, the natural form of the trusses allows for the easy accommodation of piped and cabled services.

The connection method for the upper floor joists may rely on the direct connection of the joists to the wall frame by seating the joist ends onto the head-rail of the wall panels. Alternatively, the use of a Z-section support or hanger may be adopted. In this case the joists are secured to the wall panels by the hanger, and lateral restraint is assured by the use of screwed or bolted cleats (Figure 7.43).

Roofs

The roof structure to steel framed housing can take the form of steel or timber trusses. While the use of timber trusses is quite common, the consideration will be

Figure 7.43
Connection for upper
floor joists.

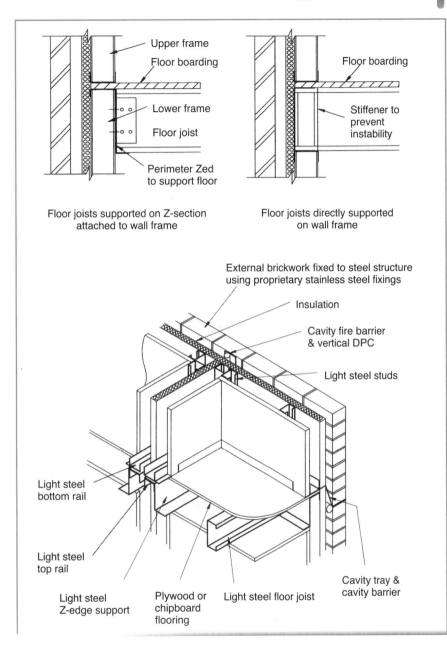

limited in this section to steel (timber roof construction is considered in Chapter 10). The strength-to-weight ratio of trusses made from C- or Z-section 'sticks' is excellent and allows for the creation of large roofs, with considerable spans between supports and with the ability to create usable space within the roof structure. As with timber roof forms, it is possible to create a wide variety of roof profiles and configurations. However, simple pitched roof forms are by far the most common (Figure 7.44). Unlike timber trussed rafters, the structural form of steel trusses is likely to take the shape of an 'open roof' (sometimes referred to as an 'attic roof'). Such construction

CASE STUDY:

UPPER FLOORS

The steel trusses that form the intermediate floors are clad with resilient floor base material in this case, as the building will be used for flats. The need to ensure acoustic separation results in the use of rigid insulation beneath the boarded surface of the floor.

The steel trusses that form the intermediate floors are visible here, supported off the open wall panels. Note the cross-bracing to the end panels and the internal wall sections.

Seen from above, the trusses that form the loadbearing elements of the floor are clearly visible. Note that, as with most types of floor joist, the trusses are positioned in accordance with the location of the lower walls and span in different directions as dictated by the shortest span.

This view of the underside of an intermediate floor shows the interaction of the trusses/joist, the internal supporting walls and the steel filling sections that resist overturning of the joists and ensure structural continuity between spaces.

Figure 7.44
Simple pitched roof.

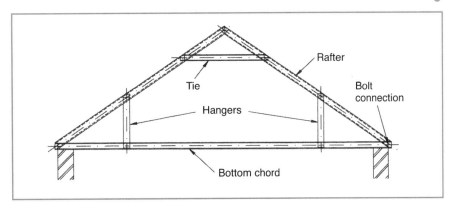

maximises the usable space in the roof and can be achieved as a result of the high bending strength of the steel members. In some instances it is necessary to incorporate C-section purlins into the structure to provide additional support for larger roofs. These function in the same manner as those used in a traditional cut roof.

One of the innovations associated with steel frame construction is the use of panelised roofs which utilise a rigid insulated board that spans between the eaves and the ridges (sometimes supported at mid-span by purlins).

The roof may also incorporate the bracing sections that assist the structure in resisting lateral loads from wind action. The bracing takes the form of lattice girders laid horizontally to link the walls to the roof structure at ceiling level. These are referred to as 'wind girders' and fulfil an essential structural requirement in ensuring that the walls and roof are connected and braced in such a way as to ensure rigidity and structural integrity in the complete structure of the frame.

See also Figures 7.45 and 7.46.

Figure 7.45
Pitched roof interior.

The interior of the pitched roof here shows the use of a structural ridge truss with inclined roof sections supported off it. This is similar to a traditional timber 'cut roof' rather than the more common trussed rafter form.

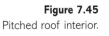

PART 3

Figure 7.46
Roof structure.

This photograph shows the structure of the roof, with internal walls and dormer sections clearly visible.

Figure 7.47
Traditional masonry cladding.

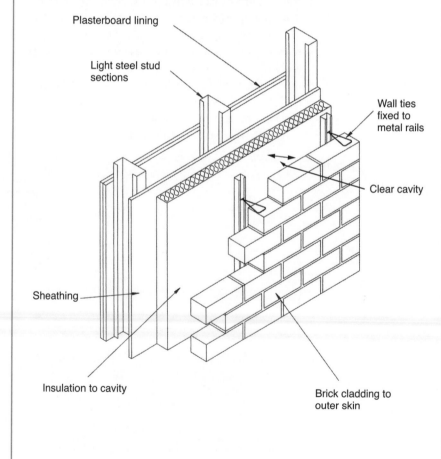

Plasterboard lining

Light steel stud sections

Wall ties fixed to metal rails

Clear cavity

Sheathing

Insulation to cavity

Brick cladding to outer skin

Claddings

As with timber frame construction, steel frame houses can be provided with a wide range of claddings to provide the weatherproof envelope of the building. By far the most common is the use of traditional masonry cladding to create a cavity wall structure in which the steel frame acts as the structural inner leaf, while the brick cladding acts as the weather-resistant outer leaf. In this form of construction the brick cladding must be restrained to prevent lateral movement using wall ties, as in the traditional cavity wall type of construction. The ties will be housed in steel channel sections that are generally secured to the steel studs of the structural frame, with the outer sheathing and insulation sandwiched between (Figure 7.47). In some cases the channels will be secured directly to the sheathing. The use of steel channels to secure the ties allows for the accommodation of vertical movement that will arise as a consequence of the different levels of thermal movement that will occur between the frame and cladding. The spacing of ties will be in accordance with structural requirements but should not be less than 2.5 ties per square metre of wall area. Additional ties will be positioned around window openings and so on.

Openings in brick cladding will be supported by independent lintels that are likely to be positioned close to the level of the intermediate floor. As with traditional cavity construction the provision of effective cavity trays, damp-proof courses and weep holes is essential to ensure the effective exclusion of moisture.

Steel frames may also be subject to the provision of rendered, boarded and tiled cladding solutions, although these are less common.

See also Figures 7.48 and 7.49.

Steel frame construction and fire

As in the case of timber frame construction, in order to control the spread of smoke and flame through the concealed voids and cavities of steel frame buildings it is necessary to install cavity barriers at key locations within the building. These must

Figure 7.48
Warm frame for a
building.

The 'warm frame' form of this building is typified by the provision of rigid insulation material to the outer face of the frame. Note the vertical steel profiles fixed to the sheathing. These provide a secure location for the wall ties that will be secured to the outer leaf or cladding

Figure 7.49
Use of rigid
insulation/sheathing
material for cladding.

The external face of the wall panels is clad with rigid insulation/sheathing material. This creates a 'warm' frame and provides for fixing of wall ties to provide lateral restraint to the brick external leaf. Once the external leaf of brickwork is erected, the building takes on a 'traditional' appearance

be capable of providing resistance to fire for at least 30 minutes. There are several methods of forming these barriers, but the most common are pre-formed tubes of mineral wool insulation material, although steel channels may be used. In addition to these cavity barriers, fire stops are required where there are a number of dwelling units within a single structural block. Fire stops are provided at the junctions of roofs and front and rear elevations with the party walls. The provision of these fire stops ensures that each individual unit forms a separate compartment capable of containing a fire for a period of 60 minutes.

CASE STUDY:

STEEL FRAME CONSTRUCTION

The foundation and floor slab are laid. In this case the building is to be built on a sloping site and the creation of a stepped profile is necessary along with retaining wall sections to cope with the differing ground levels.

The prefabricated wall and floor sections are erected with lintels/trusses over the openings in external and internal walls. In this case the floor joists are of a relatively deep, trussed form although deep C-section units are sometimes used.

The lower wall sections are completed and scaffolding is erected to allow the lift to the first floor. The external face of the wall panels is clad with rigid insulation/sheathing.

STEEL FRAME CONSTRUCTION (*continued*)

The upper storey is erected and the pitched roof structure is created. Sheathing of the wall panels is completed and the building is made weathertight with the application of the roof covering.

In this instance a rendered finish is applied to the face of the insulation material to create a durable external wall. Cladding with traditional masonry finishes is also very common.

7.5 | Openings in external walls

Introduction

- After studying this section you should appreciate the constructional detailing that may be applied when openings are formed in walls.
- You should also appreciate that there are a number of criteria to be satisfied when such openings are formed, and these may include structural, insulation, moisture and fire resistance.

Overview

The creation of openings in the external envelope of the building is essential to allow access and egress to the occupants and to allow for the provision of natural light and ventilation or the passage of services. Small openings require no special treatment, the bonding of brick walling being sufficient to allow support of the wall above. Larger openings, however, require the provision of a supporting beam or lintel to the brick/blockwork above; historically this has taken the form of an arch in many situations. Many alternative forms of lintel are available, some more suitable than others for inclusion into cavity walls. Materials commonly used for the manufacture of lintels are steel (treated to prevent corrosion) and reinforced concrete.

The forms of construction that are most commonly adopted in housing are the traditional cavity wall or, increasingly, timber or steel frame. In all of these alternatives the creation of openings must be considered carefully. In addition to the load-bearing capabilities of the lintel, a number of other factors must be noted. Of particular importance are:

- *Structural stability*: The creation of an opening in a major structural element results in the need to cater for the transfer of loads around the opening. The design of the support above such an opening and the nature of the end bearing of such a support must be carefully considered.
- *Thermal insulation*: Since the external wall provides a thermal envelope to the dwelling, the creation of openings can potentially result in excessive loss of heat. The treatment of openings must ensure that such heat loss is kept to a reasonable minimum.
- *Resistance to the passage of fire*: Since most modern houses adopt some form of cavity wall construction or timber frame construction, there is potential for the passage of smoke and flame within the cavity. Where openings are formed, there is the potential for entry of fire into the cavity. Hence it is necessary to close the cavity around openings or to provide cavity barriers to resist the passage of fire.
- *Passage of moisture*: Where cavities are closed, there is the potential for moisture to pass through the permeable masonry of the external wall to the interior of the building. Care must be taken to ensure that details are included to prevent this from occurring.

Openings in external walls: terminology

In order to consider the details involved in the creation of openings in external walls we must first become familiar with the terminology associated with such openings. There are two types of opening that must be identified. Firstly, there are the small openings or holes that are formed to facilitate the passage of pipes, ducts and other such elements through the external envelope.

These do not normally require special consideration and they are formed as required on-site. Secondly, there are the larger openings associated with windows, doors, vents and other larger items that must pass through the external wall.

In the case of the small-scale openings, the bonding of brickwork above is generally sufficient to cope with the transfer of loads around the hole. In the case of larger openings this is not the case, and we must therefore take care to ensure that the design and formation of the opening are appropriate. The careful detailing of these

Figure 7.50
Opening technology.

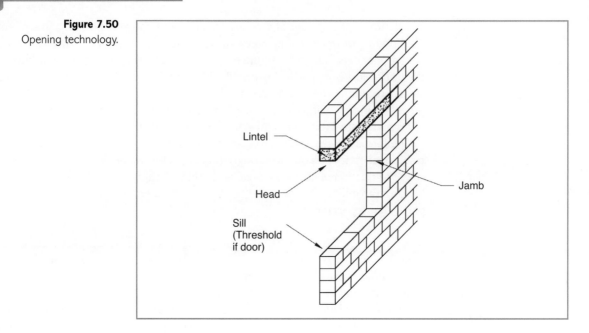

openings will ensure that there is no risk of structural failure, passage of moisture or thermal bridging. Specific details will be generated for the head (top), sill (bottom) and jambs (sides) of these openings (Figure 7.50).

Clearly, openings are necessary for access and for allowing in light and fresh air, but it should be recognised that there are limits to opening sizes in order to preserve the structural stability of the wall. Part A of the Building Regulations (Structure) outlines the relationship between opening sizes and the wall panels that remain between the openings. These wall panels are viewed almost as masonry columns between the windows. If the wall between openings were to become too slender, the wall would become unstable and the width of the window opening would need to be reduced.

Because the opening is an area of structural weakness in the wall, there is provision for extra wall ties to the vertical sides of the opening, as outlined in BS 1243. Additionally, as stability is needed at the opening, it is traditionally the case that the cavity is closed at the window or door using either the internal wall skin (usually blockwork) or the outer skin of facing brickwork. Which skin is used depends on the location of the frame within the reveal of the opening – if the frame sits towards the outside then the inner skin is used to close the cavity. If the frame sat back, this could expose the inner block skin in the reveal, and therefore the cavity would be closed with the facing brickwork.

When the cavity is closed we have effectively a solid wall, and this raises two issues – how to prevent moisture entry and how to prevent heat loss.

Figure 7.51 shows the building in of a vertical damp-proof course (DPC) at the cavity closure to isolate the brickwork from the blockwork inner skin. Where timber frames have been used these were often secured in the opening by temporary planks while the bricklayers built up the wall around the frame. In such instances it was

Figure 7.51
Vertical DPC at cavity
closures.

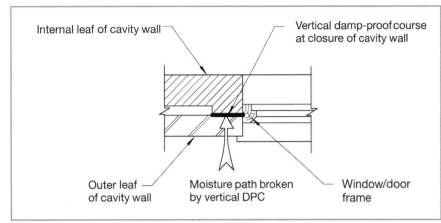

Figure 7.51
Vertical DPC at cavity
closures.

Cold bridging is the term used to describe the situation when the level of insulation of the fabric is significantly lower at a localised position than in the remainder of the fabric. In this circumstance, heat loss is concentrated in the area of lower insulative value, causing a **cold bridge** between the warm inner and cold outer air.

easy to hold the vertical DPC in position during bricklaying by nailing this to the timber frame.

Having placed the vertical DPC we have now created a barrier to moisture passage, but with still effectively a solid wall the heat loss from inside the property will tend to focus on this spot. This situation, where heat may flow more easily to the outside, is called a **cold bridge**.

Pitch polymer DPC material is now available with polystyrene pre-bonded to the back to insulate the vertical DPC. By using this detail the heat loss is substantially reduced. Figure 7.52 illustrates how this is achieved.

There are a number of ways in which the reveal area may be prevented from creating a cold bridge, and four examples are illustrated in Figure 7.53.

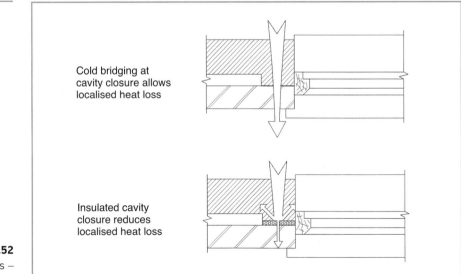

Figure 7.52
Cavity closures –
preventing a cold bridge.

Figure 7.53
Insulating to prevent cold bridges at openings in walls.

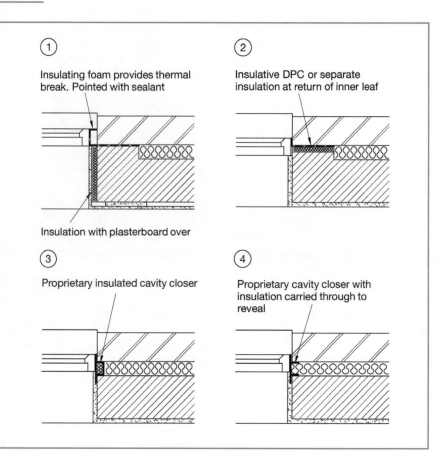

Evolution of the lintel form

Prior to the availability of concrete as a material, lintels to external and internal walls were in timber. With the advent of concrete, reinforced concrete and steel, a variety of lintel solutions emerged.

For internal walls and external walls, lintels tended to be used of the same width as the masonry of the wall in which they were located (Figure 7.54).

Where cavity external walls were used, the lintel detail had to recognise that the outer brick skin is porous and will allow moisture to run down the inner face of the external skin. Knowing that this means that moisture can accumulate on the top of the lintel, and recognising that concrete details are porous, the use of a DPC cavity tray became necessary.

Figure 7.54 shows that a **cavity tray** to encourage moisture to leave the wall externally is built into the inner skin to create the desired slope. Weep holes are generally formed in the external brick skin to allow moisture collected in the cavity to leave the wall. These may be formed by leaving vertical joints between bricks (perpends) free from mortar, or by building proprietary plastic weep holes into these joints. Two or three weep holes per opening are typical.

A **cavity tray** is a profiled impervious barrier that sits in the cavity to direct moisture to the outer face of the wall. They are often pre-formed but may be formed *in situ* using flexible DPC material. They should be installed above any horizontal bridge in the cavity to prevent moisture from entering the building.

Figure 7.54
Reinforced concrete
lintels to internal and
external walls.

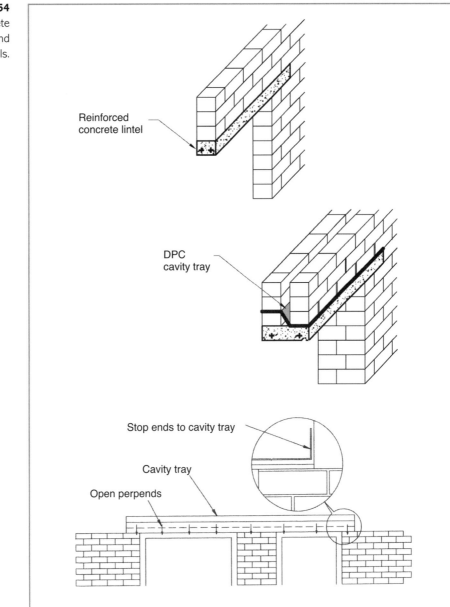

Following the use of rectangular reinforced external wall lintels the boot lintel was used, which was also able to uphold the inner and outer skins of the wall (Figure 7.55). An alternative to this was to support each wall skin separately, a reinforced lintel being used to the inner skin and a steel angle to the outer skin. This also needed a cavity tray, as shown, but this was often a preferred solution if the appearance of a lintel was to be hidden from the outside.

Figure 7.55
Alternative external wall
lintels.

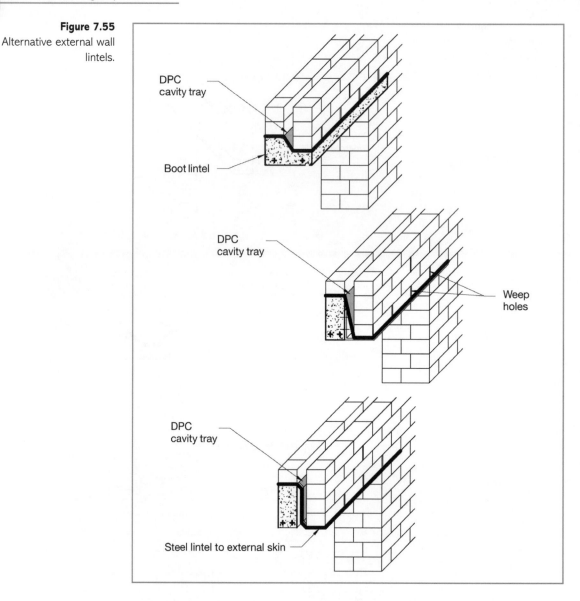

The modern solution tends to favour a steel combined lintel (Figure 7.56), this upholding both the internal and external skins. When these are used there is generally no need for the DPC cavity tray, as the steel tends to be coated against corrosion. However, thought needs to be given to cavity tray ends to hold the water collected on the lintel while it is being encouraged to leave via the weep holes.

Forms of these lintels are now typically insulated in their body to help reduce heat flow and the possibility of cold bridging. Examples of those lintels can be seen stored on-site, ready for use, in Figure 7.57.

Where the lintel is needed to an internal wall, the preferred solutions are still reinforced concrete and steel. In steel, the profile tends to be simple and corrugated for strength (Figure 7.58).

Figure 7.56
Steel combined lintels.

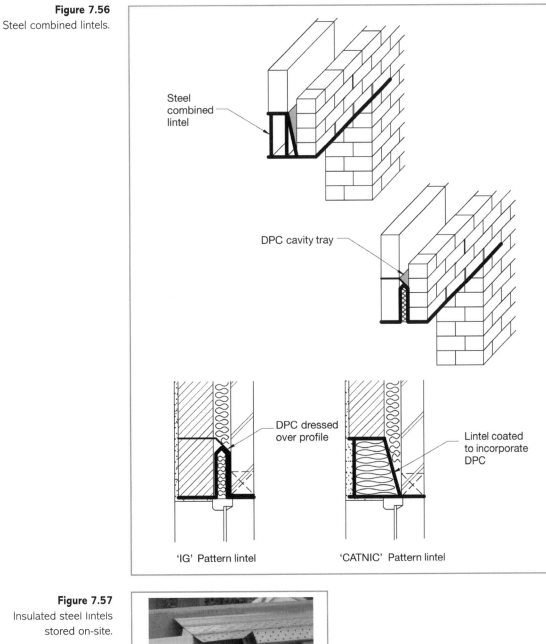

Steel combined lintel

DPC cavity tray

DPC dressed over profile

Lintel coated to incorporate DPC

'IG' Pattern lintel

'CATNIC' Pattern lintel

Figure 7.57
Insulated steel lintels stored on-site.

Figure 7.58
Steel lintels for internal walls.

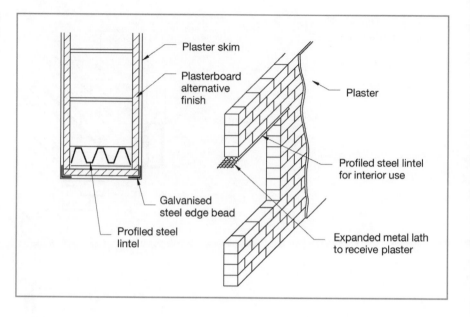

REVIEW TASKS

■ Describe *two* ways in which cold bridging is prevented in the design of cavity wall closures at wall openings.

■ Define the following wall opening terms with reference to a sketch:

   – Reveal
   – Lintel bearing
   – Cavity tray.

■ Undertake an Internet search using the term 'Lintel'. Compare and contrast the features of the various forms that you identify.

■ Visit the companion website at www.palgrave.com/engineering/riley1 to view sample outline answers to the review tasks.

COMPARATIVE STUDY: EXTERNAL WALLS

Option	Advantages	Disadvantages	When to use
Cavity walls	– Familiar technology – Good thermal performance – Good loadbearing capacity – Good moisture exclusion	– Slow – Based on 'wet trades' – Insulating cavity may compromise moisture exclusion – Difficult to achieve very high levels of insulation	– The most common form of wall in house building – Used almost as a default in speculative developments
Solid masonry	– Robust – Good structural performance	– Slow – Based on 'wet trades' – Difficult to achieve high levels of insulation without excessive thickness – May require impervious layer to ensure moisture exclusion	– Relatively uncommon in modern house building, but used in areas of high exposure because of its robustness
Timber frame	– Fast – Cheap – Can produce weatherproof enclosure quickly – Light weight allows cheaper foundations – Very good thermal performance – Removes 'wet trade' element – Reduces labour input on-site	– Poor image – Potential for deterioration if designed and/or constructed badly	– Often used on sites with low bearing pressure because of their light weight – Use where speed is of the essence or where a variety of external finishes is to be used, such as tile hanging – Becoming more popular in speculative developments
Steel frame	– Fast – Cheap – Can produce waterproof enclosure quickly – Light weight allows cheaper foundations – Very good thermal performance – Removes 'wet trade' element – Reduces labour input on-site	– Lack of familiarity	– Often used on sites with low bearing pressure because of their light weight – Use where speed is of the essence or where a variety of external finishes is to be used, such as tile hanging – Becoming more popular in speculative developments

8

Upper floors and stairs

AIMS

After studying this chapter you should be able to:

- Appreciate the functional requirements for upper floors to dwellings
- Outline the various components of upper floors
- Describe the formation of openings in floors to accommodate elements such as stairs
- Appreciate the form and function of stairs

This chapter contains the following sections:

8.1   Timber upper floors to dwellings
8.2   Stairs: design solutions and construction forms

INFO POINT

- Building Regulations Approved Document A, Structure, Section B (2004 including 2010 amendments)
- Building Regulations Approved Document K. Protection from falling, collision and impact (2013)
- BS 585: Wood stairs (1989)
- BS 5395: Stairs, ladders and walkways. Code of practice for the design, construction and maintenance of straight stairs and winders (2000)
- BS 5578: Building construction – stairs (1977)

8.1 | Timber upper floors to dwellings

Introduction

- After studying this chapter you should appreciate the performance requirements of upper floors to dwelling and the ways in which these are achieved using timber floors.
- You should understand the link between the structural performance of walls and upper floors.
- You should have developed a knowledge of the main forms of structural floor and the details associated with them.
- You should appreciate the manner in which openings are formed in upper floors and in which provision for services accommodation is made.

Overview

Although there has been little change in the layout of timber upper floors for decades, the sophistication of the materials has been somewhat refined by the details contained in the Building Regulations. These Regulations not only refer to the size and shape of components such as floor joists, but also to the quality of the materials used by reference to grade.

The primary function of a suspended upper floor must be to provide a sound, level surface, capable of supporting all dead and applied loadings over a given span. In addition, however, factors such as fire resistance, durability, thermal and sound insulation, speed of construction and the provision for services incorporation are important considerations.

Timber floor construction

Timber upper floors are the traditional solution for dwellings and have many of the advantages of timber suspended ground floors. These include providing a void space for the incorporation of services such as central heating pipes and electrical cables. They also have flexibility, warmth and aesthetic appeal.

These floors have few component parts – the joists, a boarded covering, and sometimes herringbone strutting between the joists to prevent warping.

Approved Document A of the Building Regulations (Structure) tabulates the joist sizes which may be used for a certain clear span, and generally there is more than one option which will prove satisfactory. For a particular span it is possible to use a slightly deeper joist than the alternative thicker joist (a slightly thicker joist will have less depth). The traditional thickness for floor joists is 50 mm, while depths may vary from around 175 mm to perhaps 225 mm. Other thicknesses include 40 mm and 75 mm, which will be examined later.

Joist hangers are metal fixings (normally galvanised steel) that sit into the inner leaf of the wall and support the end of the joist in a profiled 'shoe'. They provide an effective method of support without the requirement to penetrate the inner leaf with the end of the joist.

Floor joists rest on the inner skin of the external cavity wall, which in modern property is generally in blockwork (Figure 8.1). This figure also shows the alternative solution of placing the joists into pressed steel joist hangers.

When internal walls are used to support the joists, the joists have the same option of resting on the walls or fitting into **joist hangers**.

When joists are laid out for an upper floor it is commonly found that the joists run in different directions across different ground floor rooms. This is because the joists tend to run across the shortest room dimension. It should be remembered that when deciding on the joist sizes for a house the critical dimension is the longest span that is needed for the room design, as this dictates the depth of all the joists if a consistent ceiling level is to be achieved.

As joists tend to be secured at their ends, they become vulnerable to movement towards the centre and may have an inclination to buckle, as shown in Figure 8.2.

Figure 8.1
Joist ends support.

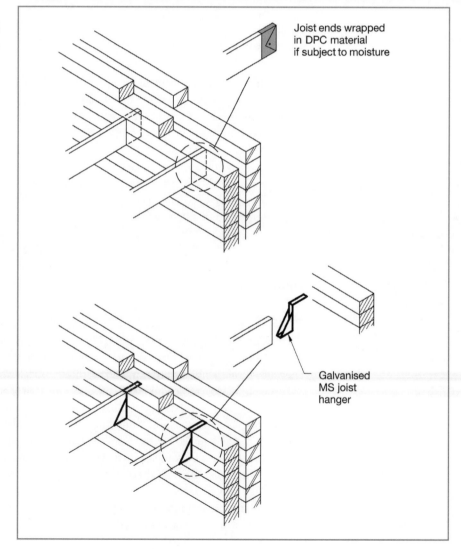

Joist ends wrapped in DPC material if subject to moisture

Galvanised MS joist hanger

Figure 8.2
Possible distortion of
joists under load.

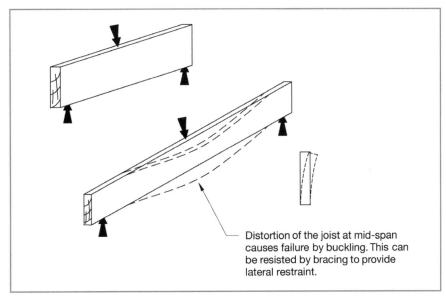

Distortion of the joist at mid-span
causes failure by buckling. This can
be resisted by bracing to provide
lateral restraint.

With the longer spans, particularly, there will be a need to stabilise the joists by either cutting short lengths of joist (noggins) to wedge between the joists (Figure 8.3) or by the provision of herringbone strutting (Figure 8.4). These are both typically located at the centre of the span.

Figure 8.3
Joist noggins for stability.

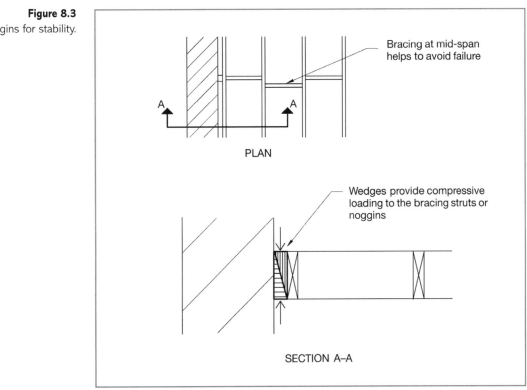

Bracing at mid-span
helps to avoid failure

A   A

PLAN

Wedges provide compressive
loading to the bracing struts or
noggins

SECTION A–A

PART 3

Figure 8.4
Herringbone strutting.

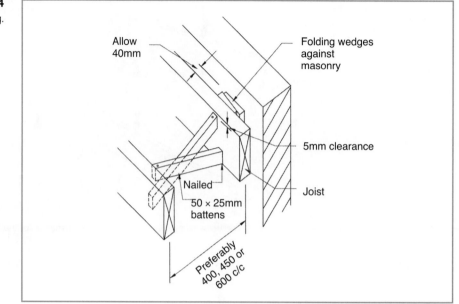

The provision of joists that can cater for relatively long spans and/or high loads requires the utilisation of large timber cross-sections. This relies on the sourcing of large timber elements from potentially mature trees. One option that allows for a more sustainable use of timber is the adoption of composite I-section joists. These are typically formed using a plywood web section with timber 'flange' sections at the top and bottom of the section. The use of such members has become common in modern house building and provides an economic, sustainable solution to structural flooring. A typical application of these units is illustrated in Figure 8.5.

Open-web joists that utilise a combination of timber flange sections and 'gang plate' steel connectors to create an open web are also increasing in use. They allow easy passage of services and offer high strength-to-weight ratios, in addition to being a sustainable materials choice.

Remember that timber is hygroscopic, which means that it has an ability to absorb and lose atmospheric moisture. As the moisture content of the timber changes, dimensional changes may occur to the joist. These will be marginal along the length of the joist, but sometimes up to 5 per cent across the grain. Moisture increases are associated with dimensional swelling, while decreases in moisture content cause shrinkage and warping.

The herringbone strutting is traditionally in timber but is now available in the form of pressed metal stage.

Coverings to upper floors include the traditional softwood tongue and grooved floorboarding and now more commonly flooring grade chipboard. Figure 8.6 shows the tongue and grooved joist of the traditional floorboarding and the kerfed joint, which is a form of tongue and grooved joint associated with sheets of chipboard.

There is a relationship between the spacing of timber floor joists and the thickness of the floorboarding applied (Figure 8.7).

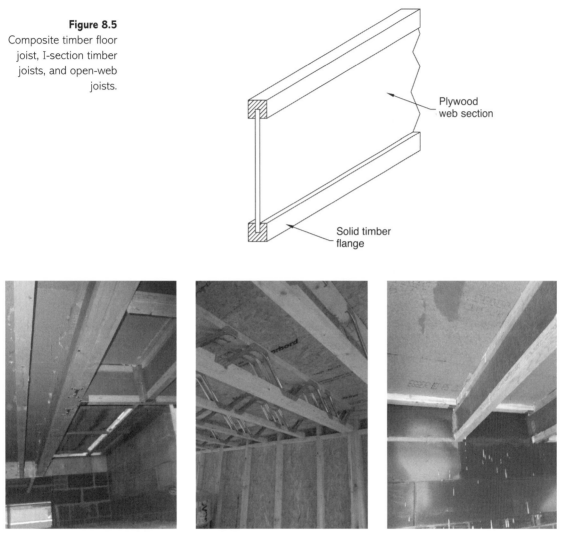

Figure 8.5
Composite timber floor joist, I-section timber joists, and open-web joists.

Plywood web section

Solid timber flange

When placing the joists to create the upper floor, the consistency of level of each individual joist is clearly an issue, as is the consistency of height above the lower floor. Figure 8.8 shows both of these factors under consideration.

Special arrangements have to be made to the floor timbers when creating an opening for the staircase. This opening is generally known as the 'stairwell'.

To form the opening in the floor, some of the floor joists have to be reduced in length, and shortened joists are termed *trimmed joists*; see Figure 8.9.

As these joists are shortened they cannot reach the supporting wall, and therefore a joist is used to pick up these ends and give them support. This *trimmer joist* is of increased thickness to be able to carry the extra load for each of the trimmed joists that it supports. In turn, the trimmer joist will require attachment at its ends to the

Figure 8.6
Joists to floorboards and chipboard sheeting.

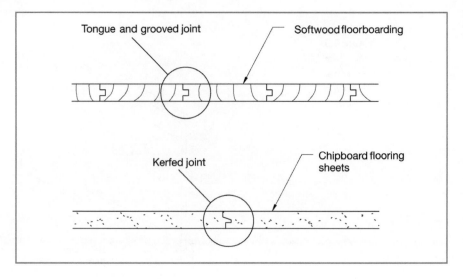

Tongue and grooved joint

Softwood floorboarding

Kerfed joint

Chipboard flooring sheets

Figure 8.7
Floorboard thickness for different joist spans.

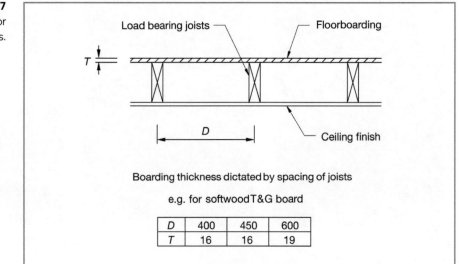

Load bearing joists

Floorboarding

T

D

Ceiling finish

Boarding thickness dictated by spacing of joists

e.g. for softwood T&G board

D	400	450	600
T	16	16	19

floor joists, which occur at each end of the stairwell opening. These trimming joists also have to be of increased thickness to take the load from the trimmer, which is carrying a number of trimmed joists.

As the opening requires a number of connections between the various joists involved, the use of metal hangers may be most appropriate (Figure 8.10).

In order to hide the sawn (not planed) timbers of the floor construction at the stairwell opening, a lining of planed softwood boarding is used which is of depth equivalent to the floor thickness. This timber board is known as an 'apron lining' and is illustrated later in the section relating to staircases.

Figure 8.8
Locating the joists –
consistency of level and
position.

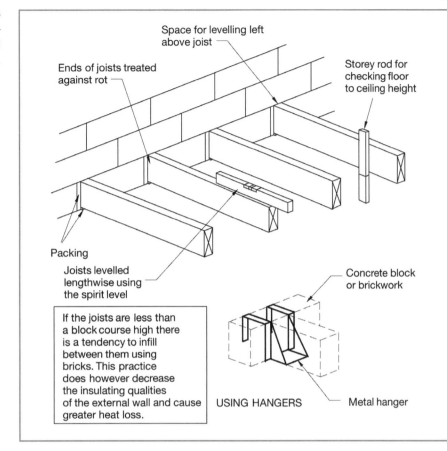

Space for levelling left
above joist

Ends of joists treated
against rot

Storey rod for
checking floor
to ceiling height

Packing

Joists levelled
lengthwise using
the spirit level

Concrete block
or brickwork

If the joists are less than
a block course high there
is a tendency to infill
between them using
bricks. This practice
does however decrease
the insulating qualities
of the external wall and cause
greater heat loss.

USING HANGERS

Metal hanger

Figure 8.9
Forming the stairwell
opening.

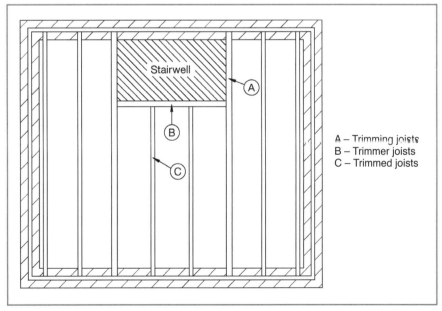

Stairwell

A

B

C

A – Trimming joists
B – Trimmer joists
C – Trimmed joists

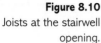

Figure 8.10
Joists at the stairwell
opening.

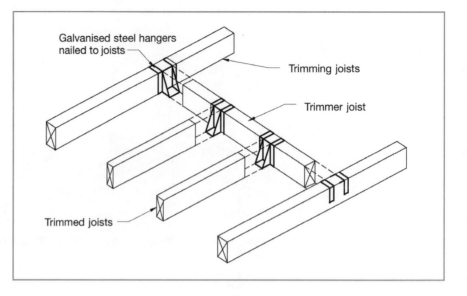

One feature of the timber upper floor is the advantage that it offers for the incorporation of pipe and cable services. Where these are to run in parallel with the spanning joists, free access is available, but it will be necessary at times to run the services perpendicular to the joist span. The consequent holing or notching of the joists needs to be limited in order to keep within the tolerances that are needed to ensure that joist strength is not affected too adversely. Figure 8.11 helps to summarise some of the restrictions that apply.

When studying the various chapters in this text, it might sometimes be forgotten that the various elements discussed often do have a major structural interrelationship. Timber upper floors provide significant assistance to the stability of the house walls through the principle of **lateral restraint**. Steel straps are attached between the house wall and the floor joists to provide the restraint, as illustrated in Figure 8.12.

Figure 8.13 also shows these straps and a general layout of some of the component parts of an upper floor in summary of the text of this section.

Lateral restraint is the term used to refer to the provision of support to a structure or an element to ensure that it resists sideways forces or movement.

Support for partitions

It is often the case that the layout of the first floor rooms in dwellings differs from the layout of the ground floor rooms. In such cases it is not possible to build partitions to the upper storey off those at ground floor level. The partitions must be supported off the suspended upper floor (Figure 8.14). This is not a problem provided that the correct approach is taken to the construction of the floor, with particular regard to the layout of the joists. Where a partition runs perpendicular to the floor joists there is no need to take any specific action in the case of a timber stud partition. This is because the load from the partition is distributed across several joists. In the case of lightweight block partitions, which are rarely used at upper floor level, it is necessary to provide a timber base plate to provide a continuous base for the blockwork. However, where the partitions run parallel to the floor joists we must

Figure 8.11
Restrictions on cuts
needed for services
installation.

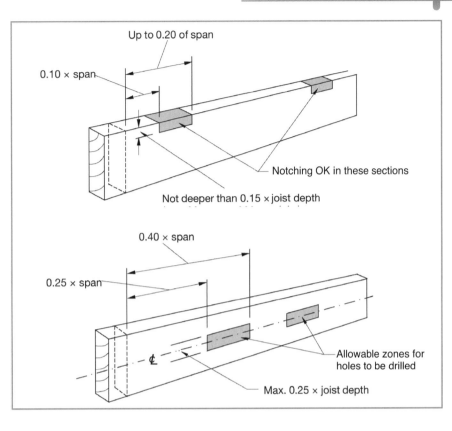

Up to 0.20 of span

0.10 × span

Notching OK in these sections

Not deeper than 0.15 × joist depth

0.40 × span

0.25 × span

¢

Allowable zones for
holes to be drilled

Max. 0.25 × joist depth

Figure 8.12
Lateral restraint straps
between wall and upper
floor.

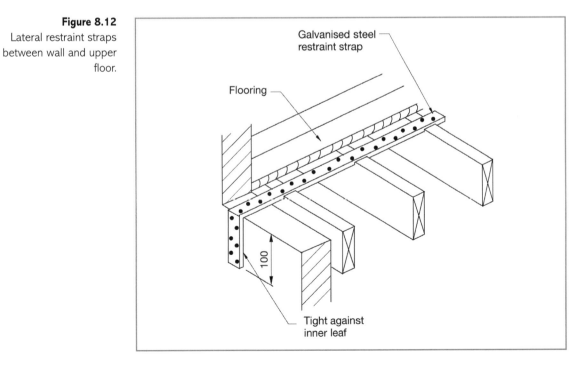

Galvanised steel
restraint strap

Flooring

100

Tight against
inner leaf

Figure 8.13
Upper floor components.

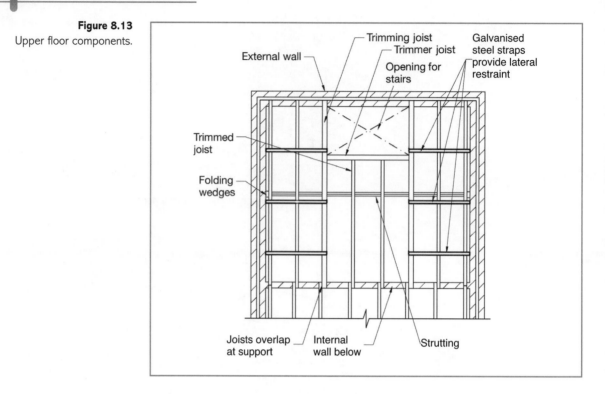

ensure that support is provided directly beneath. Since the load is concentrated, it is necessary to provide two joists bolted together or a larger section joist to cater for the additional load.

Pot and beam floors

The nature of 'pot and beam' or 'beam and block' floors has been touched on previously. To reiterate, these systems comprise a series of reinforced, often pre-stressed, profiled concrete beams, typically at 600 mm centres, between which are laid lightweight concrete blocks. The surface is then provided with a slurry screed coating enabling the application of a decorative finish.

REVIEW TASKS

■ With the aid of a sketch, describe the formation of a stairwell opening in a timber upper floor and label the name of the *three* special joists that occur at this location.

■ With reference to your own area, observe construction projects in progress – how many times are upper floors:

– Timber
– Concrete?

Why is this?

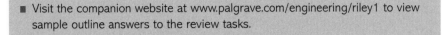

■ Visit the companion website at www.palgrave.com/engineering/riley1 to view sample outline answers to the review tasks.

Figure 8.14
Support to partitions.

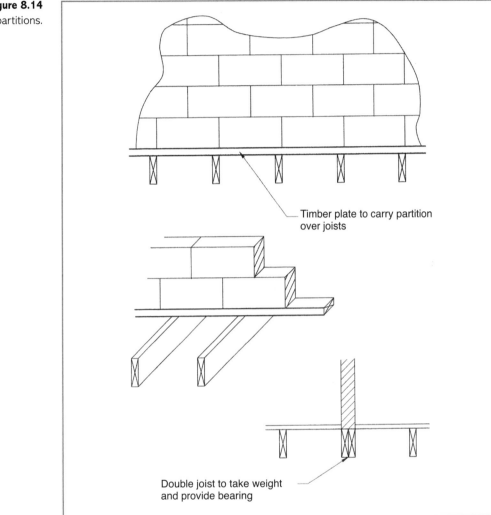

Timber plate to carry partition over joists

Double joist to take weight and provide bearing

PART 3

8.2 | Stairs: design solutions and construction forms

Introduction

■ After studying this section you should be conversant with the main design limits placed on staircases by the Building Regulations.

- You should be familiar with the terminology associated with typical domestic staircase arrangements and component parts.
- You should have an insight into the relationship between the limitations of human users and the stair design.

Staircases in timber

Timber is the main material used for the staircase of houses and the variety of types of timber stair emerges largely from the choice of whether or not to use a landing.

Figure 8.15 shows that where no landing is to be used a straight flight of steps may be adopted. Quarter- and half-turn stairs, by comparison, occur when the use of a landing allows a 90 degree or 180 degree change of direction. An older-style arrangement is where tapered winder steps provide a 90 degree change of direction while still climbing. The configuration of the stairs has a major impact upon the internal layout of a dwelling. The implication of the shape of stairs reflects upon the layout of upper and ground floors, and the length of a given stair design imposes dimensional requirements upon the building. Hence, stair designs are driven by the layout of the dwelling.

The England & Wales Building Regulations apply various limits on the dimensions of a stair in recognition of the ability of human users and to preserve safety. These

Figure 8.15
Forms of timber staircase.

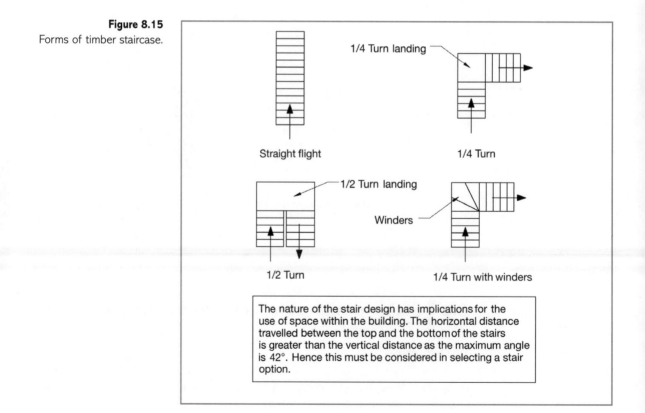

Straight flight

1/4 Turn landing

1/4 Turn

1/2 Turn landing

Winders

1/2 Turn

1/4 Turn with winders

The nature of the stair design has implications for the use of space within the building. The horizontal distance travelled between the top and the bottom of the stairs is greater than the vertical distance as the maximum angle is 42°. Hence this must be considered in selecting a stair option.

Figure 8.16
Some basic stair terminology.

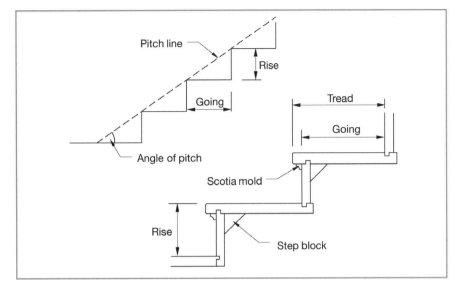

restrictions apply to many features of the stair, including design features such as the angle or slope, the steps, guard railing and headroom.

The angle or slope of the stair is termed the *pitch*, and this may not exceed 42 degrees for a private residence. In buildings of other types, a 38 degree maximum pitch applies.

Figure 8.16 shows some of the terminology of the flight of steps.

To each step there is a tread and a rise. The *going* of the step measured between the face of adjacent vertical risers tends to reflect the fact that it is unlikely that a person's foot would be placed right up to a riser. Going appears to be used in the Building Regulations in preference to the tread as a likely consequence of this fact. In the 1991 Building Regulations the step dimensions have to be such that

$$(2 \times \text{rise}) + \text{the going is to lie between 550 mm and 700 mm}$$

Additionally, the vertical rise has to lie within a certain range of dimensions to limit the height people would need to stretch each time they move forward.

From the underside, a timber stair shows that factory assembly is a fairly complicated affair. The treads and risers fit into or onto supporting *strings* at either end of the step. Figure 8.17 shows that where steps fit into these supportive strings the string is machined with housings to take the ends of the tread and riser. Such a detail involves the use of a *closed string*.

By comparison, the string may be sawn to allow the treads and risers to sit on top of the string, in which case this is called an *open* or *cut string*. Additionally, *inner strings* are fitted against a wall, while *outer strings* run between newel posts (see below).

The underside view of the steps in Figure 8.18 shows how tapered fixing wedges are driven into the tapered string housings to wedge each riser and tread firmly in position. This figure also shows the use of a central supporting timber to the steps, referred to as a *carriage*.

PART 3

Figure 8.17
Closed and open strings.

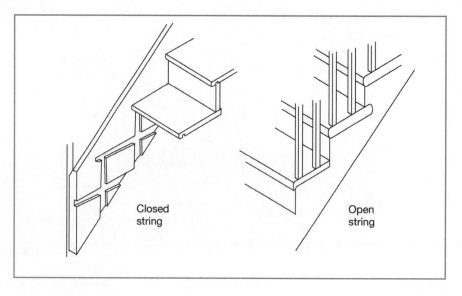

Figure 8.19 shows a variety of other basic terms which apply to the staircase. The handrails, balusters and newels collectively create the guard railing for the stair. Note the use of an apron lining at the exposed edge of the floor in the stairwell opening. Without the apron lining we would be able to see the sawn timber joists of the upper floor.

For blocks of flats and other non-dwelling situations, the staircase has to be constructed in incombustible products and tends therefore to be in either reinforced concrete or steel. These stair forms are discussed in detail in *Construction Technology 2: Industrial and Commercial Building.*

Figure 8.18
The underside of a typical flight of steps.

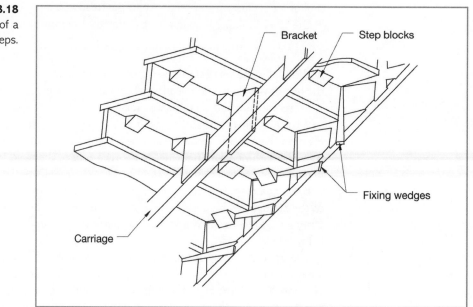

Figure 8.19
Other staircase terms.

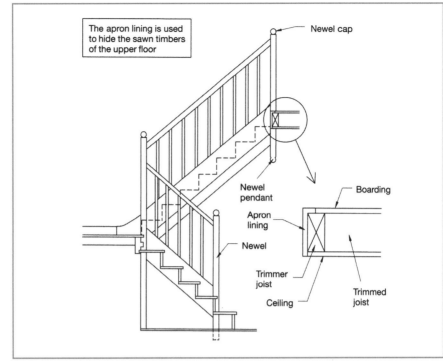

The apron lining is used to hide the sawn timbers of the upper floor

Newel cap

Newel pendant

Apron lining

Newel

Trimmer joist

Ceiling

Boarding

Trimmed joist

REVIEW TASKS

■ What feature of the stair is often used to classify the stair's arrangement?

■ Name *three* component parts of a staircase which are restricted in dimension by the Building Regulations.

■ Visit the companion website at www.palgrave.com/engineering/riley1 to view sample outline answers to the review tasks.

PART 3

COMPARATIVE STUDY: UPPER FLOORS AND STAIRS

Option	Advantages	Disadvantages	When to use
Timber suspended floor	– Familiar technology – Flexible in design – Good loadbearing capacity – Easy accommodation of services – Ready access to services	– Loadbearing limited by span and ability to obtain large joist sizes	– This form is almost always adopted in house building
Timber suspended floor (composite section)	– Sustainable – Simple technology – Flexible design – Good loadbearing capacity – Capable of long spans – Good strength-to-weight ratio – Easy accommodation of services – Ready access to services	– Sections need to be 'designed' – Slender profiles require good lateral bracing	– Increasingly used in house building
Concrete beam and block floor	– Heavy loads accommodated – Good sound insulation Excellent fire resistance	– Heavy – Services accommodation	– This form of floor is really only used where fire resistance between units is important, as in flats

9

Internal division of space: walls and partitions

AIMS

After studying this chapter you should be able to:

- Understand the functions and performance requirements of internal walls and partitions
- Appreciate the various design solutions that may be adopted for the internal division of space within dwellings, and you should have knowledge of the construction details associated with each
- Appreciate the criteria upon which selection of the various options is based
- Describe the sequence of operations involved in the formation of internal walls and partitions
- Select a suitable partition solution from a range of alternatives

This chapter contains the following sections:

9.1   Functions of partitions and selection criteria
9.2   Options for internal walls and partitions in dwellings

INFO POINT

- BS 476: Fire tests on building materials and structures (1970)
- BS 5234: Part 1: Partitions (including matching linings). Code of practice for design and installation (1992)
- BS 5268: Part 4.2: Structural use of timber. Fire resistance of timber structures. Recommendations for calculating fire resistance of timber stud walls and joisted floor constructions (1990)
- BS 8000: Part 8: Workmanship on building sites. Code of practice for plasterboard partitions and dry linings (1994)

9.1 | Functions of partitions and selection criteria

Introduction

- After studying this section you should have developed an understanding of the functions of internal walls and partitions.
- You should be able to distinguish between loadbearing and non-loadbearing forms.
- You should understand the functional requirements associated with each.
- You should be aware of the sources and nature of loads applied to internal walls and partitions.
- You should appreciate their implications relative to selection of materials and construction form.

Overview

The structure and fabric of buildings could be considered within two discrete areas: the main shell of the building and the internal elements contained within it. The shell comprises the main structural elements of the sub- and superstructures, and the internal fabric comprises non-loadbearing elements that act to subdivide spaces. Internal walls and partitions fall into the latter of these groupings. Historically the terms 'partition' and 'internal wall' were used to describe differing forms of construction. The internal wall of a building would be considered to be of more robust construction such as masonry, while the partition would be a lighter form such as timber stud. Internal walls would also generally be loadbearing, while partitions would not be expected to take structural loads. In modern construction the terms have become interchangeable and the distinction no longer exists. While it is not uncommon for the internal walls and partitions of dwellings to act as loadbearing elements, acting to support upper floors for example, the primary function of partitions that are not loadbearing is to subdivide space within the building. This is the primary function of partitions in all building types, although the nature of partitions in dwellings differs from those used in commercial and industrial building forms. The degree of flexibility required in these non-domestic types of building often necessitates the adoption of partition forms that are easy to relocate. Such forms are termed **demountable partitions**, and a wide variety of sophisticated forms have been developed in recent years.

The remainder of this section deals with the general principles of partition function, selection and form, with particular emphasis placed upon domestic alternatives.

The consideration of partitions could be undertaken with reference to two distinct constructional forms of space-dividing unit: loadbearing and non-loadbearing forms (Figure 9.1).

Loadbearing forms, in addition to dividing space, also carry some loads from the building. In most cases, this load carrying is their primary function, with the division of space being a secondary issue.

Demountable partitions are often used in commercial buildings. They are 'demountable' in that they are of a construction form that allows them to be dismantled and re-erected in different locations and configurations with relative ease.

Figure 9.1
Loadbearing and non-
loadbearing partitions.

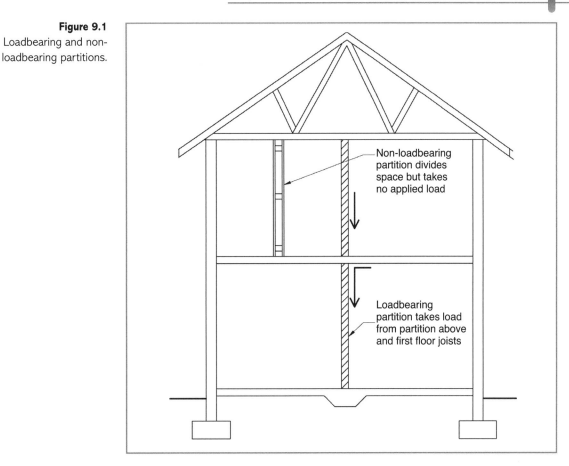

The use of non-loadbearing partitions, which carry no structural loading other than their own self-weight, is almost ubiquitous in upper floor areas of dwellings.

As with any building element, partitions are subject to a set of performance requirements that must be achieved to varying degrees in order that the element fulfils its intended function. These will take on differing levels of importance in different situations and building types. In general, however, the performance requirements for partitions may be considered as follows.

Division of space

Partitions are used primarily for the purpose of subdividing spaces within buildings. In dwellings this results in the creation of rooms or spaces that satisfy the users' needs. In non-domestic situations, partitions will be used to ensure areas of privacy, to provide visual division or simply to allocate areas of activity to individuals or operational functions. In order to achieve this it is not necessarily the case that partitions will extend to the full height of the room ceiling. Certain forms used in offices and similar situations provide only a partial separation between spaces. Such forms are

PART 3

not strictly partitions. In domestic situations the complete separation of areas is normally an essential requirement for partitions, and for internal walls.

Durability

Partitions and internal walls must be able to offer suitable levels of durability to ensure that they perform adequately for the duration of their expected life. In domestic situations the nature of the use of the rooms will impose specific demands upon the partitions. They are likely to be required to withstand loads from a variety of sources. The selection of materials and construction form must be appropriate to ensure that they provide adequate levels of durability for the lifespan of the building.

One of the most significant features of partition construction that affects their level of durability is the ability to withstand the effects of fire. The degree of **fire resistance** displayed by partitions also has a great effect upon the general level of fire safety achieved by the building. Escape routes must be maintained to allow people to exit a building in the event of fire, and the partitions and walls forming the enclosure to such routes must have adequate performance in the event of fire.

The ability of the partition to resist the passage of fire depends upon the ability of the unit to maintain its **insulation**, **integrity** and **stability** for a requisite period of time (Table 9.1). In the case of partitions or separating walls in and between dwellings, it is unusual for the required level of fire resistance to exceed 30 minutes or 60 minutes. However, in other non-domestic situations greater levels of resistance may be required.

The rate of spread of flame over the surfaces of the partition is another important factor in the control of a fire. Depending upon the use and type of the building, the partitions may be required to meet a classification between 0, when the spread of flame is almost totally inhibited, and 4, when the spread of flame is unchecked. Classifications between 1 and 4 are defined by BS 476. Although Class 0 is not defined

The **fire resistance** of partitions is defined with reference to three specific criteria. These are defined within BS 476 as:

Stability: the ability to withstand a fire without collapse

Integrity: the ability to resist fire penetration, including smoke

Insulation: the ability to resist excessive heat spread or penetration that could allow fire spread by radiation or conduction.

Table 9.1   Fire resistance of partitions.

	Form	30 minutes	60 minutes
1	Masonry: brick or concrete block	90 mm thickness	90 mm thickness
2	Stud wall with plasterboard cladding	12.5 mm boarding both sides of studs	Two layers 12.5 mm boarding both sides of studs or one layer fire-resistant boarding both sides of studs
3	Solid/layered plasterboard panels	Two layers 19 mm boarding bonded together	19 mm boarding with one layer 12.5 mm boarding bonded to each face

within the British Standard, it is referred to within the Building Regulations and is considered to provide a more strict control than Class 1.

Sound insulation

In some instances the division of space may require that sound insulation between adjacent areas is provided. This is particularly the case in situations where internal walls separate adjacent dwellings. In such situations the Building Regulations set out minimum acceptable standards for *airborne and flanking sound transmission*. In addition, it is desirable to ensure certain minimum standards of sound insulation between rooms with different uses within a single dwelling, although this is not subject to a specific requirement within the Building Regulations. The prevention of the passage of noise must be considered throughout a range of frequencies. Within dwellings this generally falls within the range 100–3150 Hz. Measurement of noise levels is sometimes undertaken with reference to dBA (acoustic decibel); this is a filtered range of frequencies that matches roughly the range of human hearing. Some partition forms may insulate well at a specific frequency while performing poorly across the entire frequency range.

The way in which the partition interacts with the surrounding structural elements is important in achieving effective sound insulation. Any holes in the partition will allow the passage of sound, and there is the possibility of sound passing around the partition if the edge details allow or if the penetration of essential building services is inadequately dealt with.

Sound absorption

In some environments the acoustic qualities of a space are very important. The level of sound that is reflected back into the space is an element that must be considered. In general it is accepted that hard surfaces reflect sound readily, and thus result in a noisier environment. Conversely, soft surfaces or those with a rough or profiled finish do not reflect sound as efficiently, thus resulting in a quieter environment. Hence the nature of the surface finish and the materials from which the unit is made are important factors in this respect.

Strength and structural stability

Partitions within dwellings are not generally loaded heavily, since they do not normally carry the loads from the main building fabric. The partitions to the ground floor area of the dwelling are likely to support the upper floor structure, and in older properties the partitions to the upper floor may provide intermediate support to the ceiling joists. In more modern construction, utilising trussed rafter roof forms, the latter is no longer needed. Whether in a domestic environment or within a commercial or industrial building, partitions are likely to be subjected to various

Figure 9.2
Loadings on partitions.

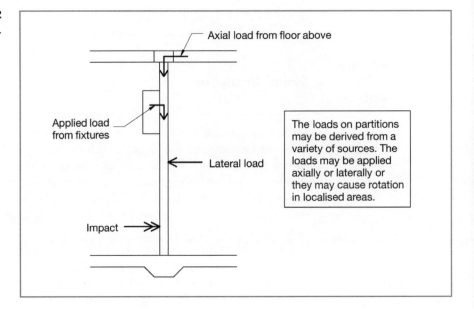

Axial load from floor above

Applied load
from fixtures

Lateral load

The loads on partitions
may be derived from a
variety of sources. The
loads may be applied
axially or laterally or
they may cause rotation
in localised areas.

Impact

forms and levels of loading and must be capable of resisting them. Forms of loading
that must be considered are (Figure 9.2):

- *Axial loading*, such as that resulting from the application of loads from roof or
  floor members resting on the top of the partition
- *Lateral loading*, such as that applied by fittings leaning on the partition, or in
  instances where people are likely to lean against the partition
- *Impact loading*, resulting from items colliding or impacting with the partition
- *Applied loadings*, such as those resulting from the positioning of fixtures and
  fittings on the face of the partition.

Appearance

In terms of the appearance of partitioning, the initial standard of internal finish
must be taken into consideration, together with any effects that this choice may
have on the provision for maintenance and ease of care during the life of the element.
In dwellings it is unlikely that the partition will be required to have any kind of pre-
applied decorative finish. It is most usual to provide the partition with a surface
coating of gypsum plaster, or to use dry finished boarding with a finishing filler
applied to the joints between panels. This allows the application of a variety of decorative treatments by the building user during its lifespan. However, in commercial
buildings it is common to require a decorative finish to be applied prior to erection
of the partition. Alternative partition forms, including glazed and panelled types,
are also adopted because of their aesthetic qualities. In general, the more sophisticated levels of finish are more costly than those of a more basic nature.

Services

Services, in the form of pipes and cables, may be encased within partitions. The type and location of such services may impose restrictions on the use of fixings and the positioning of elements that pass through the partition. Within dwellings it is unusual to allow for access to the embedded services following initial construction; however, in industrial and commercial buildings, access for the purposes of repair, maintenance and upgrading must be provided.

Although the functional requirements for partitions may at first appear to be complex, they are easily achieved in the construction of dwellings by using a range of simple and economical design alternatives. These are explored in the following section.

Domestic partitions are usually simple in form: non-loadbearing forms are most common at upper floor levels while loadbearing forms, supporting upper floor and so on, are common at ground floor level.

REVIEW TASKS

- Define *stability* and *integrity* as required by BS 476.

- How is sound reduction measured and by what scale?

- In your own home, consider the layout of partitions/internal walls. Can you identify which are loadbearing and which are not?

- What are the characteristics of each type?

- Consider the partitions to: your kitchen, bathroom and living room areas. What are the key elements required in terms of their performance?

- Visit the companion website at www.palgrave.com/engineering/riley1 to view sample outline answers to the review tasks.

9.2 | Options for internal walls and partitions in dwellings

Introduction

- After studying this section you should be familiar with the various options available for the construction of partitions within dwellings.
- You should be able to differentiate between loadbearing forms and non-loadbearing forms.
- You should be aware of the allowable forms for each.
- You should understand the implications of the form and positioning of partitions upon adjacent elements, and the required features that must be adopted in floors to support them.

- You should have a detailed understanding of the construction form and sequence for each of the available alternatives available and, given a variety of scenarios, be able to make valid judgements regarding their selection.

Overview

Partitions can be divided into a number of generic types, each with its own characteristics, advantages and disadvantages in terms of the previously noted performance requirements. They are often considered within the generic groupings of loadbearing and non-loadbearing forms. Loadbearing forms are required to withstand or transmit applied loads from intermediate floors, roof structures and other significant elements. In contrast, non-loadbearing forms are expected to carry only their own self-weight and modest applied loads from cupboards, shelves and so on. The function of non-loadbearing forms is simply that of space division. Both forms will, however, be expected to deal with loads from fixtures and fittings, such as cupboards and shelving. The extent of these loads is relatively modest in comparison to the more significant loads from floors and roofs; this is reflected in the form of the partitions selected.

Four grades of partition are identified within BS 5234: light duty used in domestic situations, medium duty for offices, heavy duty for public spaces and severe duty for industrial situations. Within dwellings the choice is restricted to relatively few alternatives. This is partly due to performance requirements, but is also closely linked to issues of cost and familiarity of traditional forms among constructors and occupiers.

Alternative construction forms

As mentioned previously, the various forms of partition can be categorised in several ways. One way of defining the types is to consider whether they are to be demountable rather than permanent. Another basis of distinction is to consider whether they are loadbearing or non-loadbearing. In dwellings, partitions are almost always permanent, although they may be loadbearing or non-loadbearing. They are considered within this section on the basis of their construction form to aid simplicity. There are many different variations available, but all can be considered within generic groupings relating to the principles of their construction. The main forms commonly found in modern dwellings include solid and hollow alternatives (Figure 9.3).

Solid partitions

The term 'solid' is taken here to mean a partition form that is free from voids in its construction. A variety of materials and design alternatives may be adopted, but the most commonly used in the construction of dwellings are masonry and laminated boards or panels.

Figure 9.3
Partition forms.

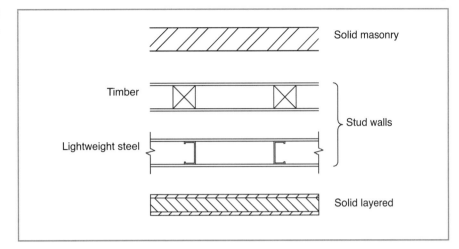

Masonry partitions

This form of partitioning is common in all building types and is typically constructed from lightweight concrete blocks or common bricks. Concrete blockwork is by far the most common material of construction, usually 100 mm thick. The use of brick for internal partitions is entirely acceptable in terms of functional performance, but this is not an economic solution. The cost of materials and the increased cost of labour resulting from the use of small brick units rather than larger blocks make this a significantly more expensive solution. In order to maximise the strength and stability of the partition, the blocks will be laid in a stretcher bond, as described in Chapter 7, and will be bonded into external walls and other internal block partitions at abutments. Partitions of this type are inherently robust, but offer limited flexibility in use. They are also much heavier than the hollow alternatives and will be required to be constructed off a thicker section of floor or to have dedicated foundations in order to support their weight (Figure 9.4).

A number of advantages are inherent in the adoption of a solid masonry partition in that it is strong, relatively cheap, easily constructed and flexible in its design and construction form (Figure 9.5). Additionally, the robust construction form allows great scope for the provision of fixtures and fittings following construction, without the need to consider any special features or additional support members. It also performs very well in terms of acoustic insulation. In some circumstances, the block or brick construction may be left **fair faced** to provide a robust, utility appearance. This approach is rare in the construction of dwellings, but is common in public buildings and areas that are not required to provide any significant level of aesthetic satisfaction.

There are, however, also a series of disadvantages which should be noted in relation to masonry constructions. Firstly, the nature of their construction is such that it introduces a **'wet trade'** into the construction process. This is an issue in terms of operational sequence and construction duration, as they are slow to erect and require a period of drying out before the application of loadings. They are also relatively inflexible in use following initial construction and are not readily altered or relocated.

Fair faced finish is the term applied to walls that are not intended to receive an application of plaster or render and are to be finished with the brick or block pointed to a good quality.

'Wet trades' are those operations that traditionally involve the use of 'wet' materials such as cement, mortar, plaster and concrete. Increasingly, 'dry' alternatives are being used to remove the time loss required while these elements dry or cure, allowing subsequent operations to take place.

Figure 9.4
Foundations to partitions.

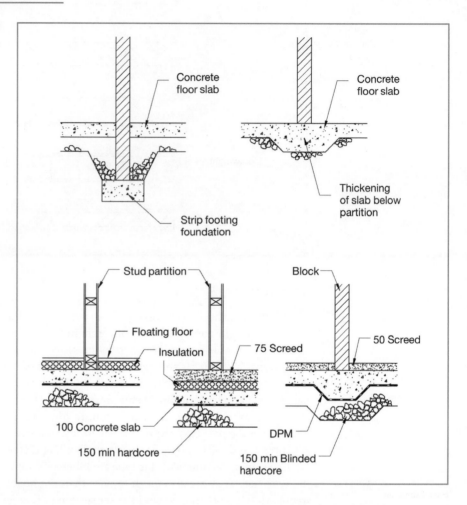

Figure 9.5
Solid partitions (lintels to openings).

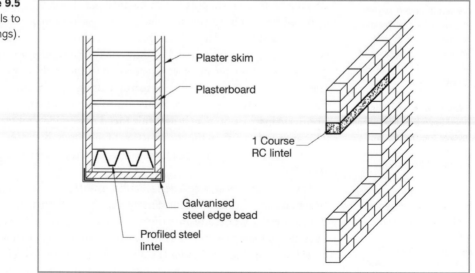

Another minor issue is that they offer limited capacity for services provision. In domestic situations, services such as electrical cables are secured to the brick or block face and are concealed by the applied surface finish. The surface finish is normally gypsum plaster.

REVIEW TASKS

- What *two* classes may internal partitions be divided into?

- What do Classes 1, 2, 3 and 4 refer to with respect to the surfaces of partitions?

- Considering a typical dwelling, where might it be essential to use a solid partition? And why?

- Visit the companion website at www.palgrave.com/engineering/riley1 to view sample outline answers to the review tasks.

Laminated and filled partitions

An alternative form of solid partition is the laminated or filled form. A number of proprietary systems are available such as 'Gyproc' laminated partitioning (Figure 9.6), which is formed by bonding layers of plasterboard to form a solid panel of the requisite thickness. Depending upon the thickness of the assembly, differing levels of fire resistance and acoustic performance are achievable. However, care must be taken when considering the application of heavy loads such as cupboards. Although not technically a solid form, the cellular filled partition is also included within this grouping. Two layers of plasterboard are bonded to a cellular core to provide a partition system that is readily assembled in panels of required thickness. However, the

Figure 9.6
Solid partitions.

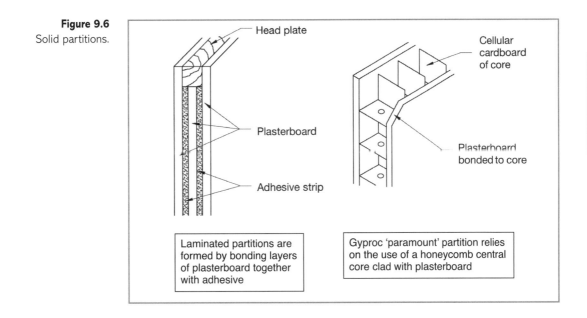

Laminated partitions are formed by bonding layers of plasterboard together with adhesive

Gyproc 'paramount' partition relies on the use of a honeycomb central core clad with plasterboard

levels of sound insulation between certain rooms required by the National House Building Council (NHBC) are difficult to achieve with this system, as the density of the overall construction does not normally provide sufficient mass to inhibit airborne sound transmission. One way of improving this situation is to provide additional layers of plasterboard cladding to either side of the construction. The incorporation of timber fixing blocks is also essential for the securing of heavy fixtures and fittings, since the partition is not in itself strong enough to support such loads without the assistance of some form of load-dissipating element.

Hollow partitions

Stud and sheet partitions

This form of partitioning comprises a series of timber or metal studs clad with a sheet material, normally plasterboard (Figure 9.7). Pre-finished systems are available, however, which do not require post-erection decorating, although these are most normally found in commercial buildings rather than dwellings. In their simplest form, as found in houses, they typically consist of a series of softwood vertical members, termed studs, with horizontal cross-pieces or noggins running between to provide bracing and fixing points for the plasterboards. This assembly forms a basic framework or carcass, which is then clad with plasterboard on each side to form the partition base. Following this, the partition surfaces may be provided with a skim coat of plaster or an application of tape to the plasterboard joints to give a smooth finish for decorating.

The assembly process of a simple timber stud partition includes the input of several building trades. The erection of the timber framing is undertaken by the joiner and is considered as a first fix joinery operation. The installation of electrical cabling and any plumbing distribution pipework to the core of the partition is then undertaken. These are first fix electrical and plumbing operations. The first fix operations are followed by the cladding and skimming of the partition, undertaken by plasterers. The process is then completed with the provision of architraves, skirting boards and electrical and plumbing fittings as part of the second fix operations for these trades. Clearly, the phasing of these operations is critical to the smooth operation of the building process. Any delays in the completion of one trade operation will have a knock-on effect on the overall pace of work.

More advanced alternatives consist of galvanised steel studs clad with pre-finished boarding, rather than plasterboard. The advantages of such forms are that they are easy to construct, the components are readily available, they are cheap, and they are flexible in design.

Their disadvantages, however, are that they can be slow to build because of the labour-intensive nature of their fabrication, and also the fact that they are constructed from sections of material which must be cut to size from large sections, so there tends to be a considerable degree of wastage. They are also somewhat inflexible in use, since once erected they do not lend themselves to dismantling and reassembly.

Figure 9.7
Stud and sheet partition.

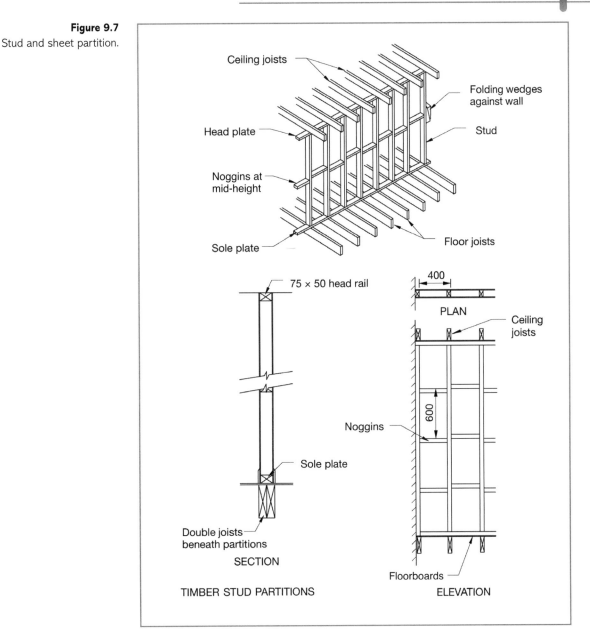

Modern rationalised construction relies on the provision of a wide variety of prefabricated components, such as windows, and manufactured elements such as plasterboard. The assembly of these individual elements is made efficient by adopting a modular approach. An example of this is seen when fixing plasterboard panels. These panels are manufactured to dimensions that are multiples of 600 mm. The placing of studs and joists at 600 mm centres reduces the need to cut panels and therefore minimises wastage.

REVIEW TASKS

■ What name applies to partitions whose body is formed largely in layers of plasterboard?

■ Where would you find noggin pieces and what is their function?

■ Describe the difference between sound insulation and sound absorption.

■ Visit the companion website at www.palgrave.com/engineering/riley1 to view sample outline answers to the review tasks.

COMPARATIVE STUDY: PARTITIONS

Option	Advantages	Disadvantages	When to use
Solid: masonry	– Robust with good loadbearing capacity – Simple and familiar technology – Provides good base for fixtures and fittings – Can be left unplastered in some environments – Good sound insulation – Inherently fire-resistant	– Heavy – Slow to erect – Based on 'wet trade' – Inflexible in use/difficult to alter after construction	– Ground floor partitions carrying upper floor loads – In areas subject to heavy use or likely to suffer damage – Between areas requiring good sound insulation – When attempting to create fire compartment
Solid: laminated/ filled	– Good fire resistance – Simple technology and easy to erect – Fast – Dry installation	– Need to order in advance – Components heavy and unwieldy	– Often used to accelerate the building process – Ground and upper floor locations – Used where a dry construction process is desired
Hollow: stud and sheet	– Simple and familiar technology – Cheap – Can be a dry process	– Heavy fixtures need support members – Inflexible in use/difficult to alter after construction	– Almost ubiquitous in house building for non-loadbearing situations

10

Roof: structure and coverings

AIMS

After studying this chapter you should be able to:

- Describe the functional performance to be provided by roofs
- Appreciate that when dealing with roofs we tend to divide the constructional detailing into roof structure and roof covering
- Describe the different basic methods of forming pitched and flat roof structures
- Describe the different forms of roof covering available for pitched and flat roofs
- Describe the various methods used for the collection of rainwater from roofs
- Understand the problems created by condensation and appreciate the need for good levels of thermal insulation and ventilation

This chapter contains the following sections:

PART 3

INFO POINT

- Building Regulations Approved Document A, Structure, Appendix A (2004 including 2010 amendments)
- BS 402: Specification for clay plain roofing tiles and fittings (1990)
- BS EN 490: Concrete roofing tiles and fittings. Product specifications (2012)

- BS 680: Specification for roofing slates (1994)
- BS 747: Specification for roofing felts (1999)
- BS EN 1304: Clay roofing tiles for discontinuous laying (1998)
- BS EN 1443: Chimneys. General rquirements (2003)
- BS 4978: Specification for visual grading of softwood (1996)
- BS 5268: Structural use of timber. Code of practice for trussed rafter roofs (2006)
- BS 5534: Code of practice for slating and tiling (2003)
- BS 6399: Part 3: Loading for buildings. Code of practice for imposed roof loads (1998)
- BRE Digest 299: Dry rot:its recognition and control (1993)
- BRE Digest 307: Identifying damage by wood boring insects (1992)
- BRE Digest 327: Insecticidal treatments against wood boring insects (1992)
- BRE Digest 345: Wet rots: recognition and control (1989)
- BRE Digest 346: The assessment of wind loads (1992)
- *BFCMA Guide to choosing and using flues and chimneys for domestic gas burning appliances* (2007)

10.1 | Functions of roofs and selection criteria

Introduction

- After studying this section you should be able to understand the functional performance to be provided by the roof element.
- You should understand the development of tension and compression forces in the roof.
- You should appreciate its insulating qualities.

Overview

The functional performance criteria required of external primary elements to the building typically include:

- Strength
- Weather resistance
- Durability
- Insulation
- Aesthetics.

Strength

Houses are generally built with loadbearing external wall structures where the inner skin of the external cavity wall (or timber/steel frame) carries the load from the roof and upper floor. Roof loads are transmitted through the inner leaf or frame from the point of connection of the roof to the wall. This connection is key to the satisfactory performance of the roof/wall combination. It should be remembered that when dealing with roofs the normal procedure is to divide this element into two parts: the *structure* and the *covering*. Both of these component parts require strength, and we will review each in turn.

The use of timber is almost ubiquitous for roof structures, and the strength of the timber structure of a roof is dependent on the strength of the material – the species of timber, the number of natural features such as knots, and the dimensions (particularly with reference to the unsupported span). Timber is graded by Part A of the Building Regulations Approved Document A, Section 1B, Table 1, in accordance with BS 4978.

Table 2 of the above document also shows the relationship between the strength required and the pitch of the roof.

The way in which these timbers are then arranged to carry the load over the required span is also an issue when examining strength. The sections that follow will examine the various roof design layouts.

Most construction components are designed to withstand two principal forces: *compression* (*squashing forces*) and *tension* (*stretching forces*). Whether the roof structure is traditional or uses prefabricated trussed rafters, both of these forces will have to be successfully absorbed. The characteristic strength developed by a roof design needs not only to be able to carry the gravitational loads of the timbers and roof covering the span, but also live loads which may arise from wind action, rain or snow. (See BS 6399 and BRE Digest 346.)

Weather resistance

Weather resistance has always been the main function of a roof, and over the years many coverings have emerged to achieve this function: thatch, timber shingles, slates and tiles for example.

Recognising that this is a vital function, the Building Regulations Part C, clause C4 reviews the need to prevent the passage of moisture to the inside of the building. There are many features of the roof purposely designed to resist moisture entry, and each of these will be examined when reviewing the constructional details which apply.

A key issue in resistance to weather entry is the slope or pitch of the roof.

Figure 10.1 shows some of the typical roof designs that have evolved over the years. The slope or pitch has particular significance when using slate or tile roof

Figure 10.1
Roof designs.

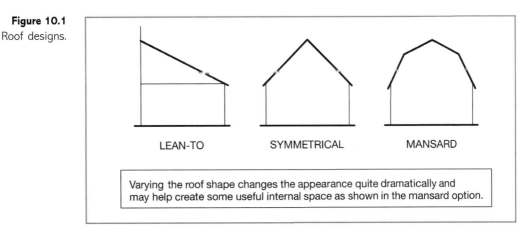

LEAN-TO   SYMMETRICAL   MANSARD

Varying the roof shape changes the appearance quite dramatically and may help create some useful internal space as shown in the mansard option.

Figure 10.2
Wind action and roof shape.

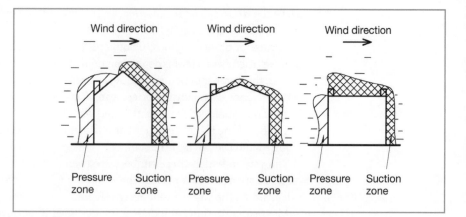

coverings, as the shallower the pitch the greater the likelihood of rain being driven underneath the covering by wind action: see Figure 10.2.

As will be appreciated later, there is a secondary line of defence to such rain entry in the form of **sarking felt** laid over the roof structure and below the covering. If rain reaches this felt it will be carried down to the rainwater gutter, as the sarking felt terminates at gutter level. The shape of the roof is also significant in terms of the resultant patterns of positive and negative pressure arising from wind action. The negative pressure or suction experienced by the roof tends to do far more damage in strong winds than does positive pressure. The extent of both pressures is dramatically influenced by roof slope (pitch).

Sarking felt is an impervious sheet material that is laid beneath the tiles or slates of a roof to provide a secondary layer of defence against weather penetration.

Durability

The ability of the roof covering to be durable has been linked with resistance to rain and to wind action. However, today atmospheric pollution may also be a factor in the equation. Sulphur dioxide in the atmosphere is the source of weak sulphuric acid and *acid rain* – just one of the airborne pollutants to be resisted.

Durability of the timber structure of a roof relates to its strength in many ways and to the sectional size of timbers originally used when forming the roof. The structure will need to resist both gravitational and live loads over the design span between supportive walls. Sometimes the development of rot and insect attack may have a considerable influence on the long-term future of the roof (see BRE Digests 299 and 345 for fungal rot and BRE Digests 307 and 327 for insect attack).

Most domestic timber roof structures are softwood, and this material is now generally treated with preservatives to resist damage over the building life.

Insulation

If we examine the pattern of heat loss from houses we can see that the roof is of major importance in thermal resistance. As warm air naturally rises, the tendency is for heat to be lost through the roof, and this fact is reflected in the current Building

Regulations Part L and Approved Document L, which impose the greatest need for insulation on the roof element. Currently the U value for a roof is 0.2 W/m^2 K (one-fifth of a watt from each square metre of roof when a one-degree temperature difference exists between the inside and outside of the roof). This requirement can be compared with the U value for the wall, which is currently 0.3 W/m^2 K (allowing almost 50 per cent more heat loss as the roof or half a watt per square metre heat loss for each degree difference in temperature between inside and outside). The shift away from defined U values to the Target Emission Rate approach means that designers have flexibility regarding the detail of specific components.

Proposals are in hand to make the insulation standards even more onerous in the near future.

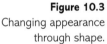

The **dBA** (acoustic decibel) **scale** is a filtered range of frequencies that matches roughly the range of human hearing.

Insulation, though, means not only insulation of heat but also against sound entry. The **dBA scale** is the means by which we measure sound levels, and the roof can be quite effective in making decibel (dB) reductions to create an acceptable internal environment. High noise generators, such as aircraft, may still prove problematic however, if the property is located on a flight path to or from an airport, or if the property is close to a major transport route such as a motorway.

Aesthetics

The shape of a roof is generally dictated by the layout of the house walls which support it and which have a major influence on the final overall appearance of the building. The use of hipped ends, for example, may have a significant effect on the appearance, as shown in Figure 10.3.

Figure 10.3
Changing appearance through shape.

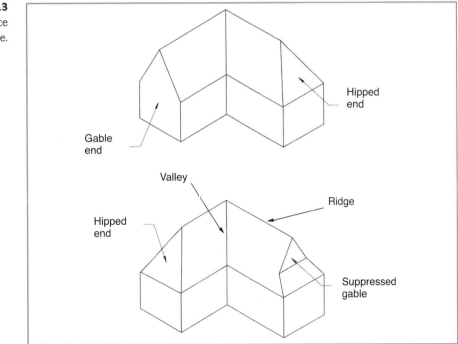

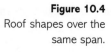

Figure 10.4
Roof shapes over the
same span.

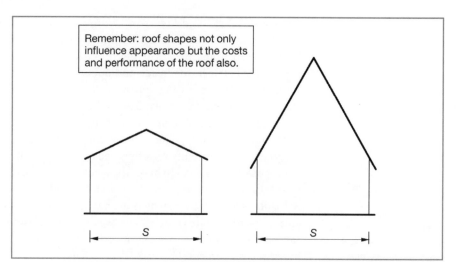

Remember: roof shapes not only
influence appearance but the costs
and performance of the roof also.

The slope or pitch of the roof will also have an impact on aesthetics.

Different visual effects will arise from use of different roof coverings and their colour, texture, shape (Figure 10.4), size and laying pattern.

The more complicated the plan layout of the walls of the building, the more complicated the roof structure solution. Easy solutions will be found where there are two gable walls and a straight run of trussed rafters in between (see Figure 10.11). When the building is other than rectangular in plan the roof shape moves accordingly and two hipped ends or a hip one end with a gable end at the other (as shown earlier) may be common solutions.

It should be remembered that the more complicated the shape, the more costly the roof solution.

REVIEW TASKS

■ Why is the roof of particular concern with respect to thermal insulation? How does the roof insulation compare to the insulation value required for external walls?

■ What is one of the biggest influences in the overall aesthetic impact of a roof design?

■ In the area where you live, try to identify a range of different roof shapes and pitches. Can you identify trends in the shape and covering? What is the difference between the covering to shallow roofs compared to that of steeper roofs?

■ Visit the companion website at www.palgrave.com/engineering/riley1 to view sample outline answers to the review tasks.

10.2 | Pitched roof forms

Introduction

- After studying this section you should be able to describe the difference between traditional rafter and purlin roofs and modern trussed rafter roof structures.
- You should be aware of roof terminology.
- You should appreciate the link between the design of a pitched roof and the span between supporting walls.
- You should be aware of particular details to ensure roof stability.

Overview

A key ingredient in the design of a roof structure is the span between supporting walls. This is illustrated by consideration of the development of traditional pitched roof designs: designs for constructing the roof out of individual pieces of timber without prefabrication (Figure 10.5).

For small spans between supporting walls, such as in the case of a garage, pairs of sloping rafters are joined at the apex of the roof by the use of a ridge board. This detail, referred to as a 'couple roof', is inherently weak structurally as the tendency is for the roof to collapse in the centre, pushing out the tops of the walls. The span for which this roof form remains stable is very limited.

A much more structurally sound detail is created by introducing a horizontal timber between the pairs of rafters, as in the case of a collar roof solution. This collar acts as a tie (a member in tension), preventing the rafters from moving outwards and creating a roof structure which, viewed in elevation, takes an A-profile shape.

The other variation is what tends to be applied for domestic housing, namely the closed couple roof. Here horizontal ceiling joists run over the timber wallplate (on top of the inner skin of the cavity wall) and connect to the foot of each pair of rafters. This triangulates the roof structure and provides significant restraint to the outward horizontal spread of the rafter. This ceiling joist (also a tie in tension) provides the means of support for the plasterboard sheets which will later become the ceiling to the room below. The ceiling joist is the main tie for the roof and is absolutely vital in restricting the natural tendency of the rafters to move outwards, displacing the top of the wall. Without the ceiling joist, the deformation experienced would be termed 'roof spread'.

Whether the roof is traditionally constructed or constructed using prefabricated trussed rafters, there are a number of terms that generally apply to pitched roofs and these are illustrated in Figure 10.6.

Rafter and purlin pitched roofs

Rafter and purlin roofs (sometimes referred to as 'cut roofs') are the traditional form of pitched roof used in housing. The roof solution, as said earlier, is constructed

PART 3

Figure 10.5

Variations of pitched roof structure.

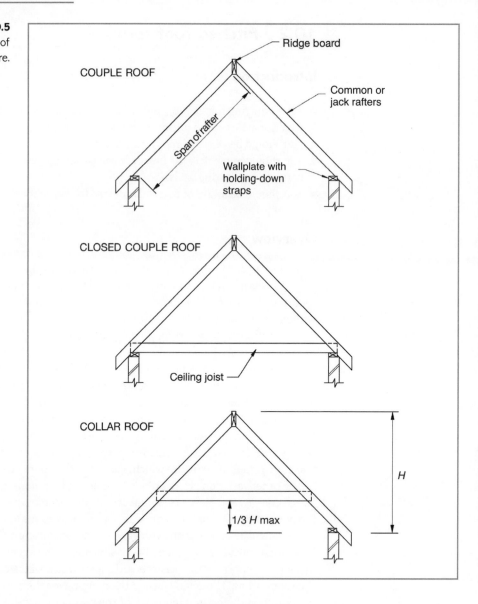

from individual pieces of timber without any prefabrication. Figure 10.7 shows the terminology typically associated with this roof form.

Where rafters are full length and extend from the eaves to the ridge, they are termed *common rafters*. In any location where rafters need to be cut to join a hip rafter or a valley rafter, these shortened versions are termed *jack rafters* (Figure 10.8).

Clearly, the function of the rafter is to carry and support the weight of the roof covering, whether it is in slates or in tiles. As the rafter is to transfer its load onto the external wall as well as the purlin, there is a need to spread the heavy point load from each rafter along the wall top in order to keep the wall top stable. The spread is achieved using a timber wallplate, which is usually 100 × 75 mm in section and

Figure 10.6
Pitched roof terminology.

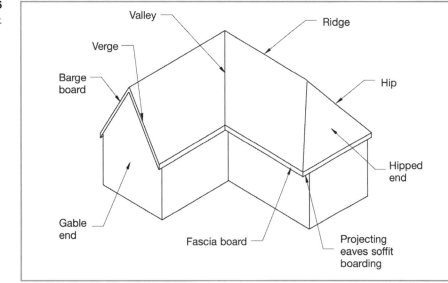

Figure 10.7
The rafter and purlin roof structure.

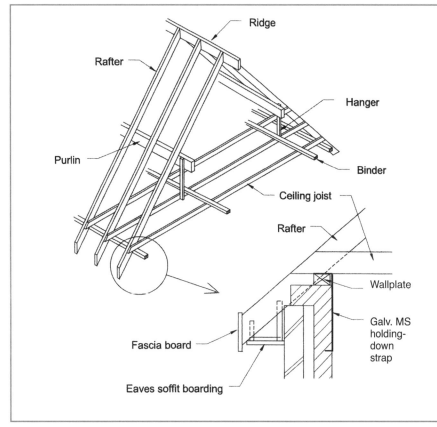

Figure 10.8
Common rafters and jack rafters plus hip rafter-supported purlin.

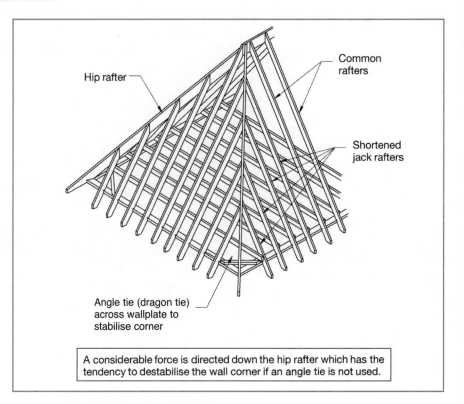

Hip rafter

Common rafters

Shortened jack rafters

Angle tie (dragon tie) across wallplate to stabilise corner

A considerable force is directed down the hip rafter which has the tendency to destabilise the wall corner if an angle tie is not used.

Figure 10.9
Load spread through bonding masonry wall.

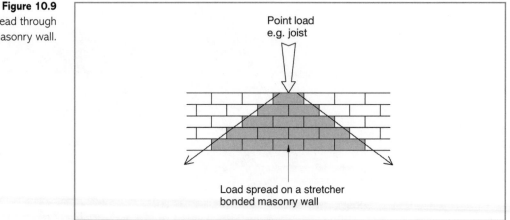

Point load e.g. joist

Load spread on a stretcher bonded masonry wall

placed over the inner skin of the cavity wall. The wall plate has another function in that it helps anchor the rafter end by allowing it to be notched over the plate. It also supports the ceiling joists which are attached to the rafters at this level, spreading the loads from the joists along the wall as well.

The gravitational and live loads (such as wind) carried by the roof are spread quite efficiently at the point of load transfer to the wall. Further spread of load is achieved by laying the blockwork of the inner skin of the cavity wall in stretcher bond. Figure

PART 3

Figure 10.10
Purlin location.

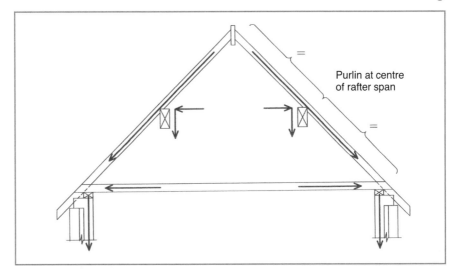

Purlin at centre
of rafter span

10.9 shows how the loads move down through the wall and this of course means that by the time these loads reach the foundation they have considerably reduced in magnitude by the effect of spreading.

The heaviest timber in a traditional rafter and purlin roof is the purlin itself. This is a member located below the rafter to carry some of the rafter load (Figure 10.10). If only one purlin line is used, the rafter will be supported midway between the wallplate and the ridge. Sometimes two lines of purlins are used which divide the rafter span between the wallplate and the ridge approximately into thirds.

Support for the purlin is achieved at a gable end by building the purlin into the brickwork. Here the purlin was traditionally carried through the external wall and used to support the underside of the roof overhang. Today this support is generally provided by using a **gable ladder**.

A **gable ladder** is a ladder shaped assembly of rafters that allows a projection of the roof at the verge.

Where we have a hipped end to the roof or where the purlin meets a valley, support for the purlin is provided by the hip rafter or valley rafter used at these points (Figure 10.11). This places significant responsibility on these special rafters to carry the considerable loads taken by the purlin, and as a result some of the hip and valley rafters are of substantial depth (typically >> 300 mm). One of the features of the rafter and purlin roof is that the loads are catered for by creating a 'triangle' structure in which the vertical and oblique loads are catered for by the connections at wallplate and purlin. In practice the rafters lean against each other at the ridge board with equal and opposite force. For this reason the ridge board is not designed to take any vertical loading, but rather acts merely as a connection point for the upper ends of the rafters.

Dormer is the term used to refer to a window projection within the roof profile that provides for additional headroom in the roofspace and facilitates the provision of natural light and ventilation.

One advantage of the rafter and purlin pitched roof solution is that the roof space can be used for storage or, provided suitable sections of joist are used, for accommodation. In such circumstances it is normal to incorporate a **dormer** into the roof structure. The pitch or slope of the roof will dictate how much roofspace is available. In previous years, when slates or plain tiles have been used, the pitch has ranged

Figure 10.11
Hip rafter support to
purlin.

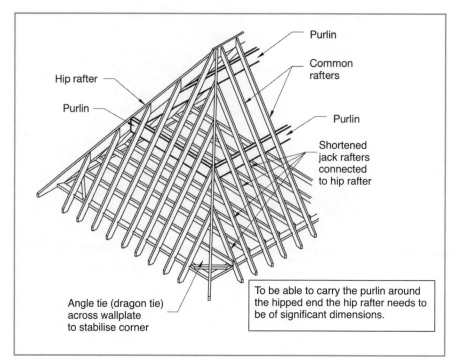

Purlin

Common
rafters

Hip rafter

Purlin

Purlin

Shortened
jack rafters
connected
to hip rafter

Angle tie (dragon tie)
across wallplate
to stabilise corner

To be able to carry the purlin around
the hipped end the hip rafter needs to
be of significant dimensions.

from around 35 degrees to 45 degrees, and this has provided a good roofspace. It should also be remembered at this point that there is a link between the pitch or slope of the roof and the nature of roof covering to be employed. Where the pitch is shallow, perhaps for example only 22 degrees, the roof covering needs to be of an interlocking artificial variety to resist penetration of the covering by wind-driven rain (see the next section).

To date, the pitched roofs examined have been symmetrically pitched, with rafters each side of a ridge. When the rafter runs up to a wall at the top we form a simple *lean-to* or *pent* detail. The eaves are identical to symmetrical pitch eaves, and the same comments apply regarding the size of pitch and the nature of the roof covering (see Figure 10.1).

A **gang nail plate** is a method of securing general members of the trussed rafter together using a galvanised plate incorporating numerous nail-like projections. This is pressed into adjacent timbers by laying it across the joints and using high pressure impression jigs to assemble in the factory.

Trussed rafter pitched roofs

Many of the reports concerned with the efficiency of the construction industry (for instance, Sir John Egan's report for the DETR: *Rethinking Construction: The Report of the Construction Task Force*, July 1998) have recommended prefabrication as a method of improving efficiency by reducing labour intensity on-site. One of the elements which has particularly changed as a result of prefabrication is the timber pitched roof structure. A 'trussed rafter' is the title that is given to the detail which combines rafter and ceiling joist (Figure 10.12).

The trussed rafter eliminates the heavy purlin by replacing it with internal bracing, and eliminates the ridge board by using a **gang nail plate** as a result of factory rather

Figure 10.12
A simple form of trussed
rafter.

Figure 10.12
A simple form of trussed
rafter.

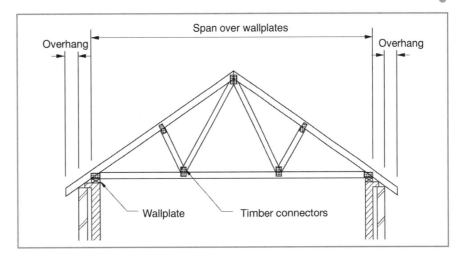

than site assembly. As shown in Figure 10.13, the basic triangle of component parts is provided by the two rafter sections and the horizontal ceiling joist which, as with the traditional rafter and purlin roof, is still a tie member in tension. Once the wallplate is positioned on the top of the wall, the prefabricated trussed rafters are lifted into place, and fixed at 600 mm.

Figure 10.13
Component parts of the
trussed rafter.

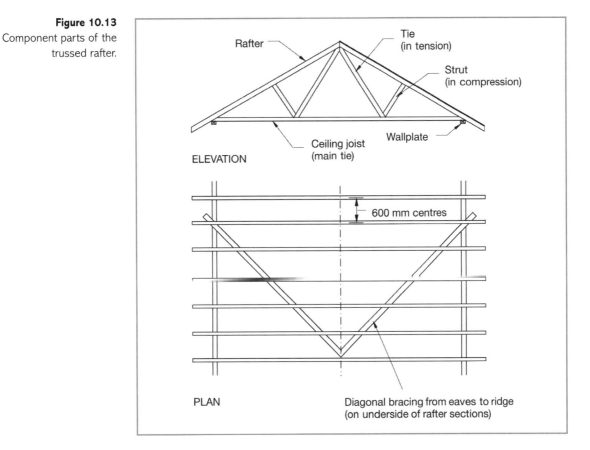

PART 3

Figure 10.14
Trussed rafter variations.

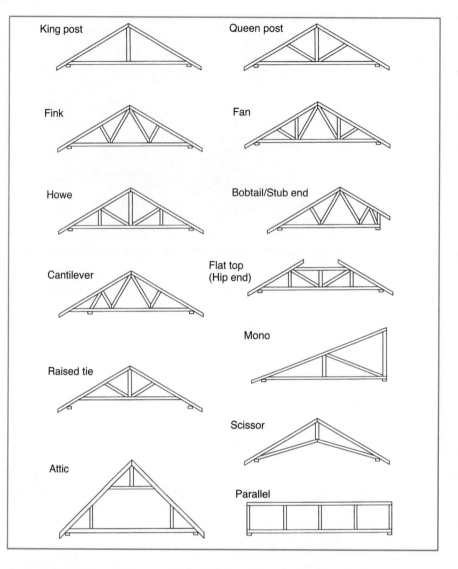

With simple roof designs, the basic roof structure can be placed by two operatives in a matter of hours, and is far superior to the rafter and purlin roof in terms of time for completion. Of course the use of internal bracing to this detail does restrict movement inside the roofspace. However, trussed rafter forms have been developed in a wide variety of configurations, including those that allow use of the roofspace.

If we examine the function of the internal braces, the two internal timbers which join the rafters at the apex of the detail are always members in tension (ties), as they tend to support the horizontal ceiling joist and stop it sagging. By contrast, the shorter members give support to the rafters and stop them sagging by providing a propping function. These members replace the supporting role of the purlin, and as such take compression loads (struts).

The most common profiles for the trussed rafter, widely used in housing, are shown in Figure 10.14. Where the internal members of the trussed rafter form a W

Figure 10.15
Trussed rafters stored
on-site.

(two ties and two struts), this is typically known as a *Fink* arrangement. As the span increases between supporting walls, so does the number of internal members to the trussed rafter. Owing to the relatively slender section of the trusses it is essential that they are transported and stored on-site in a vertical position, with suitable support, otherwise they may suffer from twisting. Examples of trusses stored on-site are shown in Figure 10.15.

It should be remembered that in this situation the two internal members that meet at the apex of the detail are always ties in tension supporting the horizontal ceiling joist, while all of the other members are struts in compression supporting the rafter.

Once the basic trussed rafters for the roof have been assembled and spaced at typically 600 mm, other timbers will be seen running horizontally between the trussed rafters providing anchorage to space them apart. Alternatively, pre-positioned metal truss clips may be used for location and anchorage purposes (Figure 10.16).

Figure 10.16
Truss clip anchorage to
the wallplate.

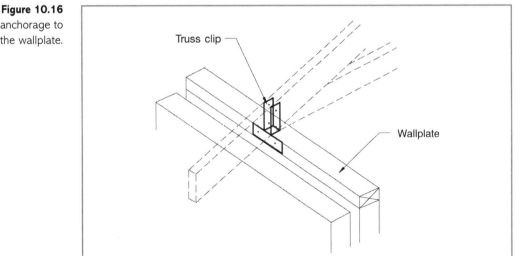

PART 3

Figure 10.17
Longitudinal bracing and
diagonal wind bracing.

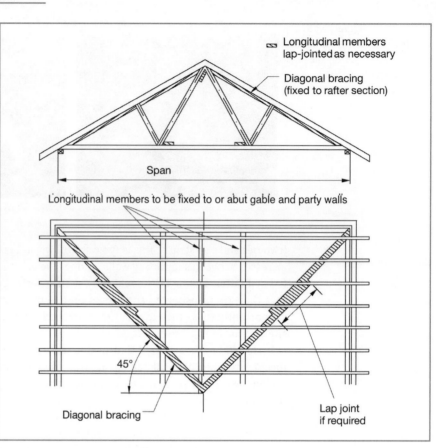

Figure 10.17 Longitudinal bracing and diagonal wind bracing.

Another key feature of the system is the timbers used to resist collapse by wind pressures, and this bracing runs diagonally from the wallplate attached to the underside of the rafter and up to the apex of the roof; see Figure 10.17.

Figure 10.18 shows two photographs of a rafter roof during the process of construction.

To assist in transportation, the pitch angle of trussed rafters is often quite shallow, and this dictates that interlocking tile roof coverings are used as the roof covering. Pitches in the low twenties of degrees may be expected.

At a gable wall, it is often the case that the roof is designed to project beyond the wall for appearance benefits. When this is so, a gable ladder is constructed to cantilever the roof beyond the wall, as shown in Figure 10.19.

If it is the intention to locate the main cold water storage tank within the roof-space, the weight of this needs to be spread over a number of trussed rafters by use of a timber platform (Figure 10.20). Remember that 1 litre of water weighs 1 kg.

On simple gable-ended roofs the trussed rafter detailing is simple, and of course this means minimal cost. However, where the roof turns direction or is hipped, more complicated and more expensive detailing will be needed. One drawback of the prefabricated roof structure is that it tends to require careful and special consideration where the roof changes direction. Where roof shapes are simple (straight roof

Figure 10.18
Trussed rafter roof during
construction.

between gables) the prefabricated roof can save significant amounts of money compared with the traditional solution.

If the wall to the house is to project into a feature such as a bay and the pitched roof is to continue over this detail, then special trussed rafters of varying size or strength may be needed: see Figure 10.21. When hipped ends occur there are two options: use special trussed rafters of variable size or form the detail out of individual rafters rather than use prefabricated components.

Figure 10.19
Gable ladder extension of
roof over gable wall.

PART 3

Figure 10.20
Support to cold water
storage tank.

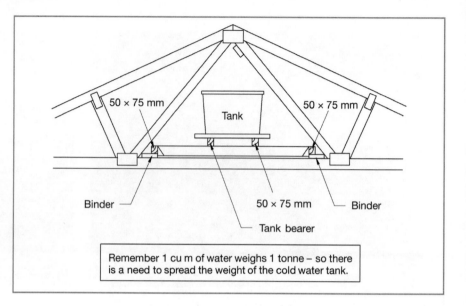

50 × 75 mm

Tank

50 × 75 mm

Binder

50 × 75 mm

Binder

Tank bearer

Remember 1 cu m of water weighs 1 tonne – so there
is a need to spread the weight of the cold water tank.

Figure 10.21
Trussed rafter junctions
between roof slopes.

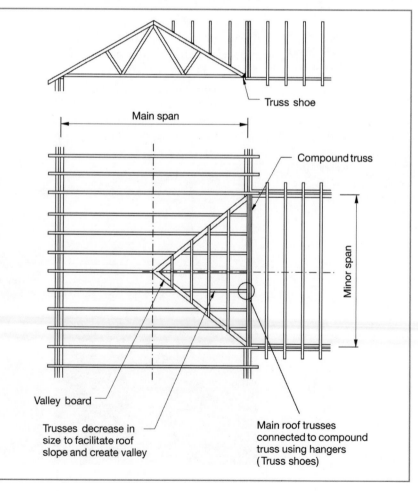

Truss shoe

Main span

Compound truss

Minor span

Valley board

Trusses decrease in
size to facilitate roof
slope and create valley

Main roof trusses
connected to compound
truss using hangers
(Truss shoes)

Figure 10.22
Roof dormers.

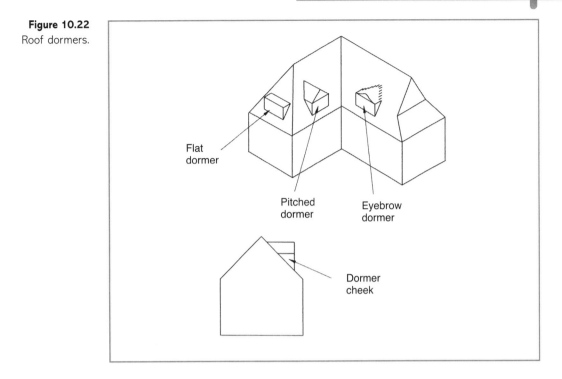

Roof dormers

It is quite common to incorporate 'dormer' windows into pitched roofs to dwellings. The dormer is a small protrusion from the roof that incorporates a window and allows the localised mere use of increased ceiling height internally. Several configurations are possible but the most common are simple flat or pitched roof forms (Figure 10.22).

REVIEW TASKS

- Where would you find a 'jack rafter'?

- Where would a purlin be located?

- Name *two* internal component parts of a trussed rafter.

- Where would you find diagonal bracing of a roof structure and what is its function?

- Undertake an Internet search for the term 'trussed rafter'. Using this search, locate manufacturers' websites and compare and contrast the various truss forms that are available.

- In the area where you live, identify housing construction sites. Can you locate any developments that opt for rafter and purlin construction? Why might this be?

- Visit the companion website at www.palgrave.com/engineering/riley1 to view sample outline answers to the review tasks.

10.3 | Pitched roof coverings

Introduction

- After studying this section you should be able to appreciate the differences in the laying technique required for different forms of roof covering.
- You should understand the link that exists between the slope or pitch of the roof and the roof covering.
- You should be familiar with the options that are available for pitched roof coverings.
- You should understand the criteria associated with the selection of different coverings.

Overview

There is quite an extensive choice when selecting coverings for pitched roofs, largely as a result of the various shapes and colours of the manufactured options.

Broadly, the range of options includes slates, plain tiles and interlocking tiles. These will now be discussed in more detail.

The common feature of pitched roof coverings is that they are based on the principle of layers of overlapping 'plates' of material in the form of flat slates or tiles. The performance of the roof will be affected by the pitch of the roof, the extent of overlap and the nature of the material used.

More modern materials use greater levels of sophistication in ensuring effective weatherproofing.

Slates

Slate is a natural material excavated from a quarry and cut to thickness and size. This is a naturally hard and durable material, but within the slates quarried there are variations in hardness which cause them to be cut to different thicknesses and these are given classifications such as *bests* or *seconds*. The process of quarrying and processing slate relies on exploiting the natural lamination of the material to give thin layers. 'Bests' are usually thinner.

These are some of the largest forms of this style of covering used on pitched roofs, traditionally identified in imperial sizes such as 12 × 24 inches (approximately 300 × 600 mm). A length of twice the width is typical.

As they are natural and not artificial materials, each slate has to be holed for nailing to the timber roof battens, and two holes per slate will be located either towards the top (head nailed slates) or towards the centre (centre nailed slates). The latter technique is usually employed for strength, where severe exposure to wind action is expected (Figure 10.23).

Figure 10.23
Head nailed and centre
nailed slates.

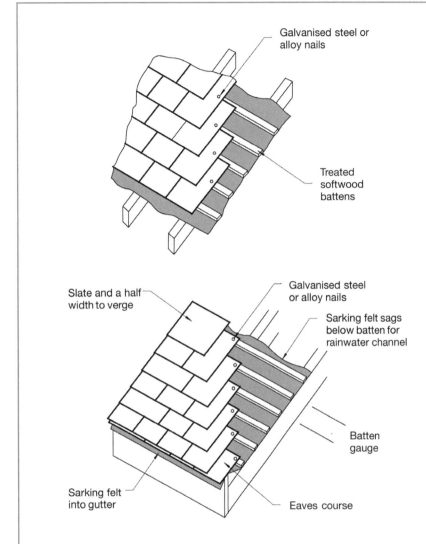

The illustrations also show the need for bonding when laying and, to start off the bond, special slate-and-a-half widths are provided at the verge (roof edge at a gable wall) on every other course. The reason for bonding, to prevent rain entry, is highlighted in Figure 10.24.

Note with respect to the illustrations that the gauge is the centre-to-centre spacing of the fixing battens and that there is typically a double thickness of slate over the entire roof, with a triple thickness at the lap position. Modern slate roofs have a sarking felt underlay which intercepts any rain entry and carries any penetrating rain down to the rainwater gutter for disposal. Prior to the use of sarking felt, the slate would be *torched* internally, torching being a mortar-like application to seal the top of each slate course inside the roofspace.

Figure 10.24
Preventing rainwater
entry with butt jointed
roof coverings – multiple
layer coverings.

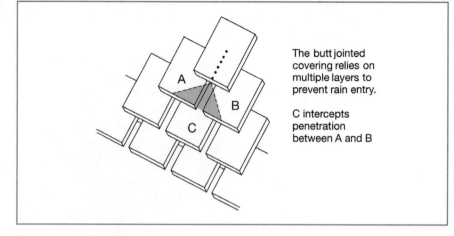

The butt jointed
covering relies on
multiple layers to
prevent rain entry.

C intercepts
penetration
between A and B

Plain tiles

Unlike slates, plain tiles are factory-manufactured units. They generally use one of
two materials, either clay or concrete. As a manufactured unit they are formed in
such a way as to possess features that would be desirable but impossible in slates –
a camber in the length to resist capillary movement of water between tiles, and two
nibs at the top for hooking over the fixing timber battens. Each tile is also pre-holed
with two holes in the factory ready for nailing. However, it is often the case that
nailing does not occur to every individual tile. In normal exposure situations it is
common to nail perhaps only every fourth course, as the weight of tile upon tile
holds the tiles in position. In exposed situations every tile may be nailed to resist
wind uplift.

Plain tiles are small units measuring typically 265 mm × 165 mm × 12 mm in
thickness and as such many are required to complete one square metre of roof
covering. The nature of the material is such that the process of covering a roof is
highly labour intensive and relatively costly. Plain tiles tend to be either body
coloured if clay, or surfaced coloured if concrete. In the latter case, colours tend to
fade with time and the loss of colour has a significant effect on overall appearance.

As with slates, plain tiles must be bonded from course to course and a one-and-a-
half width tile will be necessary to start off the bond at the edge of the covering on
every other course. Figure 10.25 shows a section through a plain tiled roof covering
at the ridge position and this shows terminology shared with slates: *gauge* (the
centre-to-centre spacing of the fixing battens), *head lap* (the overlap between tiles)
and the fact that over the covering generally there will be at least a double thickness
of tile material with a triple thickness where the head lap is measured.

At eaves level the connection to the supportive roof structure and sarking felt
below the roof covering is important. The felt is the secondary line of defence to
water entry and carries any rainwater that penetrates the tiles down to the rainwater
gutter. Note that a special short eaves course provided at gutter position again maintains the double thickness of tiles over the full area of the roof and is needed for the

Figure 10.25
Plain tiles at ridge level.

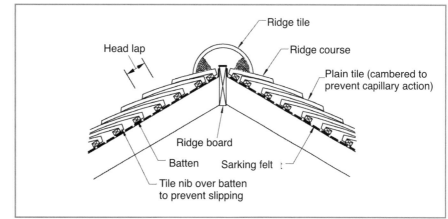

reasons illustrated in Figure 10.26. Note also that the roofspace is ventilated through the eaves area, either through ventilation grilles in the eaves soffit boarding or through vent ducts located below the roof covering at eaves level. Cross-ventilation of the roofspace should be achieved, as the eaves to each side of the roof are ventilated. The Building Regulations require that this ventilation gap varies with roof pitch. However, provided the gap is a minimum of 25 mm, any roof pitch can be used.

At the verge of the roof covering an undercloak is provided (Figure 10.27), traditionally in the form of slates. One of the functions of the undercloak is to tilt the roof edge to hold water on the roof slope.

When a roof changes direction there needs to be continuity of the barrier to protect from rainwater entry, so where a valley is formed there are a number of

Figure 10.26
Ventilation of roofspace through the eaves soffit boarding.

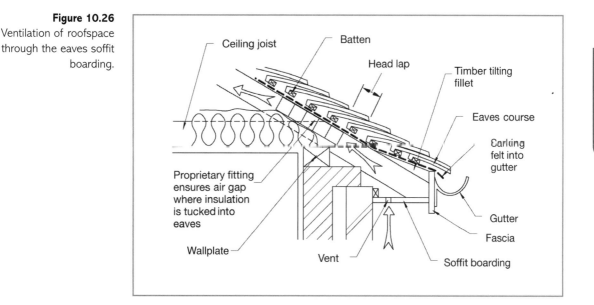

PART 3

Figure 10.27
Undercloak course at the
roof verge.

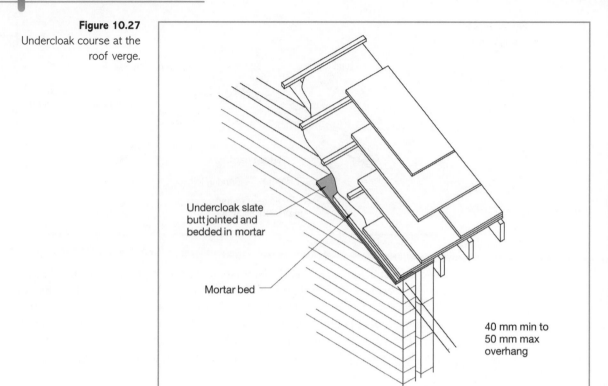

Undercloak slate
butt jointed and
bedded in mortar

Mortar bed

40 mm min to
50 mm max
overhang

Figure 10.28
Lead lining to the valley
gutter.

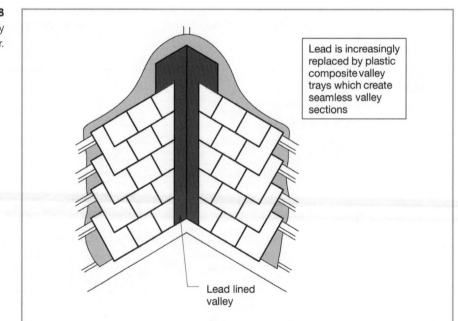

Lead is increasingly
replaced by plastic
composite valley
trays which create
seamless valley
sections

Lead lined
valley

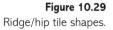

Figure 10.29
Ridge/hip tile shapes.

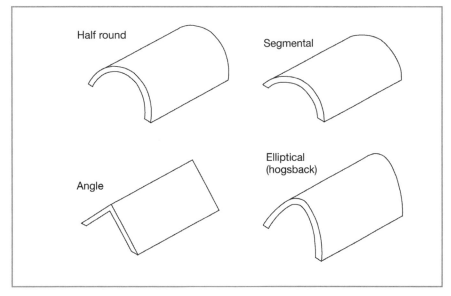

optional details that may be employed. Traditionally here we would place boarding of about 200 mm width on both rafter slopes at the valley to act as support for a lead lining to the valley (Figure 10.28), finishing the tiling or slating so as to leave the open lead-lining channel.

Alternatively, we can obtain purpose-made valley tiles in order to be able to continue the tiled covering around the change in slopes or sweep the tiling on a curve around the valley.

Lead, the traditional valley lining, has a limited although fairly long life, and today more modern materials can be obtained for even longer life in the form of plastics. These rolls of valley lining are dished to hold the water and have grips to each edge to hold the valley line against the sarking felt, and are simply cut to the required length. Timber valley boards will be used to provide support.

At the ridge of the roof, a gap is naturally formed as the roof tiling finishes to each roof slope, and this gap is covered with ridge tiles which are usually supplied with the roof tiling as an accessory and in order to have continuity of roof material. The ridge tiling may be in a number of different sectional profiles, as illustrated in Figure 10.29.

When a hip is formed in the roof, ridge tiles are extended from the ridge down the hip and are now termed in accordance with their new function: that is, 'hip' tiles rather than 'ridge' tiles. At the foot of the hip there could be a tendency for the hip tiles to be encouraged to move under gravity should they break their bond with the mortar which is used to bed them in place. To resist this movement a hip iron is screwed to the hip rafter, creating a barrier to possible slip (Figure 10.30).

As an alternative to these planted-on hip tiles, special bonnet tiles (Figure 10.31) may be used to bond into the general tiled roof slope. These are nailed into position and then mortar filled, as shown in the detail.

PART 3

Figure 10.30
Hip iron to secure the hip tiles.

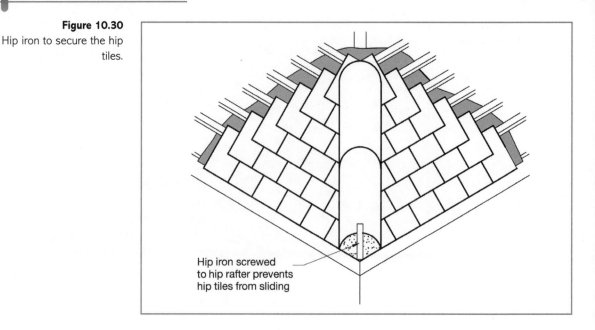

Hip iron screwed
to hip rafter prevents
hip tiles from sliding

Figure 10.31
Bonnet hip tiles.

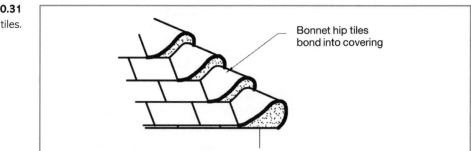

Bonnet hip tiles
bond into covering

Interlocking tiles

The formation of a plain tiled roof covering involves many hundreds of small units, and consequently the speed of making this covering is slow. A more modern equivalent is the much larger interlocking tile.

The other major advantage of this tile is its resistance to rainwater entry, caused by the fact that, as the name suggests, tiles physically interlock together through a series of projections and corresponding grooves.

These characteristics of this tile also change the laying pattern. With many of the proprietary forms of interlocking tile there is no need to bond when laying, and as a result this tile may be laid in straight rows up the pitch of the roof. Among other things, this means that there is no need for special tile-and-a-half widths at verges, and over the general area of roof there is only a single thickness of material except at the headlap, where the bottom of one tile overlaps with the head of another. Figure 10.32 shows sections through this type of tiling close to the ridge and eaves details.

Figure 10.32
Interlocking tile details.

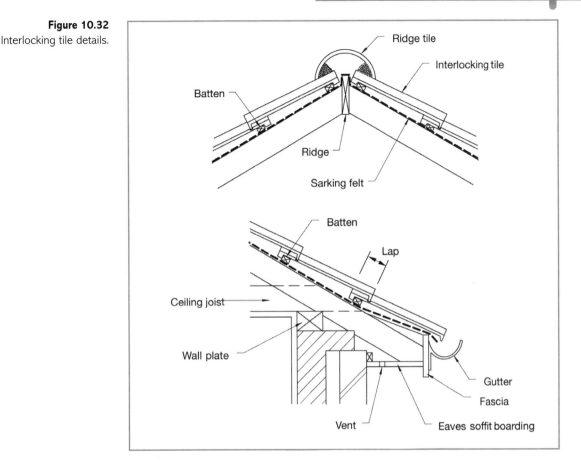

Figure 10.32
Interlocking tile details.

As there is only a simple lap arrangement with this form of tile, they are referred to as a *single lap* covering.

Dry roof details

The traditional method of finishing details to the verge and ridge of pitched roofs relies on the use of cement mortar to set and seal the ridge tiles and to seal the exposed edge of the tiles at the verge. These are important details as they ensure that the roof does not suffer from moisture penetration at these locations. In addition these details guard against damage caused by wind acting upon the roof, which could dislodge tiles. The process of applying mortar at these locations creates the need for a skilled 'wet trade' to finish the roof covering. In recent years this has been almost entirely superseded in new house construction by the use of 'dry' details. These rely on the use of purpose-made verge and ridge trims, formed in plastic or PVCu, that are secured and sealed without the use of mortar.

In the case of the verge detail, a series of individual verge trims or 'caps' may be secured to the edge of each tile course, secured by secret fixing to metal clips that

Figure 10.33
Dry roof details.

are nailed to the ends of the roof battens. Figure 10.33 shows the clips clearly visible at the verge. These clips act as a securing mechanism for the edges of the verge tiles and also provide a secure restraint for the verge caps. In areas that are prone to high winds the caps may be nailed directly to the roof battens to provide additional security of fixing. A less common alternative is the use of a 'linear dry verge', which utilises a continuous profile that is fixed along the fill length of the verge in a single section.

Dry ridge detailing is achieved by utilising overlapping or interlocking cementitious or PVCu ridge tiles that are secured using secret fix clips or by screwing to a timber ridge section, They are then sealed at the ends using pre-formed edge capping that match the dry verge caps.

REVIEW TASKS

■ Describe *two* differences in the laying principles of interlocking tiling and slate roof coverings.

■ What is the function of sarking felt and where does this material terminate?

■ When would centre nailed slates be used in preference to head nailed?

■ Visit the companion website at www.palgrave.com/engineering/riley1 to view sample outline answers to the review tasks.

10.4 | Flat roof forms

Introduction

■ After studying this section you should be able to understand the different forms that a flat roof may take.
■ You should appreciate the importance of details to prevent condensation within the roof.
■ You should understand the differing performance characteristics of different forms.

Overview

'Flat' roof construction is something of a misnomer, as this form of roof does require a slope of at least 1 in 80 to clear rainwater. The ways in which this is achieved will be discussed later.

As a construction detail this form of roof is to be avoided if at all possible, because of its established poor performance record in respect of maintenance and weathertightness. As a rough guide, one might expect to have at least three to four times the trouble-free period with a pitched roof compared with a flat detail, and today the economical trussed rafter means that the cost of a tiled pitched roof is often comparable. However, modern polymeric and other materials have enhanced the performance of flat roofs greatly.

Problems tend to develop as a result of the deterioration of the covering externally or from condensation of water vapour within the fabric of the roof as warm moist internal air meets cold surfaces in close proximity to the outside temperatures.

In an attempt to improve the performance of the flat roof with respect to internal condensation problems, construction detailing of flat roofs has changed in recent years: ventilation of the cold deck solution, which represents the original flat roof form, has been improved.

There are three main forms of flat roof which are distinguishable by the position of the thermal insulation relative to other parts of the detail (particularly the deck). Figure 10.34 illustrates the three main forms of flat roof which are possible, but it should be remembered that in most cases, and particularly in housing, it is the cold deck detail which is most widely used. The warm deck and inverted roof forms perform better in terms of prevention of condensation and durability. However, their use is most widespread in construction forms other than housing. Therefore they are discussed in detail in *Construction Technology 2: Industrial and Commercial Building*.

Cold deck flat roof

As shown in Figure 10.34, this roof is so called because of the fact that the thermal insulation is located below the deck. This is a good position for intercepting the heat which is rising from the building, but it also isolates the timber deck, leaving it cold and providing a surface against which moist warm air may condense. A vapour barrier, such as a polythene sheet, tends to be used (always on the warm side of the insulation) to prevent moisture-laden air reaching the cold deck, but this barrier is quite difficult to achieve successfully.

Timber, which is usually used for roof structures, is a hygroscopic material, which means that it can absorb and lose moisture. As it gains moisture it expands, and as it loses moisture it shrinks and often distorts. If condensation causes moisture absorption into the timber we may reach the moisture content level where fungal growth is encouraged, which can have a devastating effect on structural integrity. Cross-ventilation from side to side of a flat roof by eaves soffit vents is far more difficult to achieve successfully than it is in pitched roofs, as mentioned earlier. The Building Regulations suggest that the ventilation gap should be the equivalent to a continuous vent 25 mm wide for flat roofs with a span up to 5 m, and to 30 mm for spans of 5–10 m.

The main body of the flat roof, as illustrated in Figure 10.35, consists of timber joists generally set at 400 mm centres to suit the dimensions of the plasterboard which will be attached to the underside (soffit) of the joists to form the ceiling finish. The Building Regulations Approved Document A (Structure) should be consulted regarding the grade and section of timber to be used for the clear spans of the design. As with the timber pitched roof, it is normal to place a timber wallplate over the

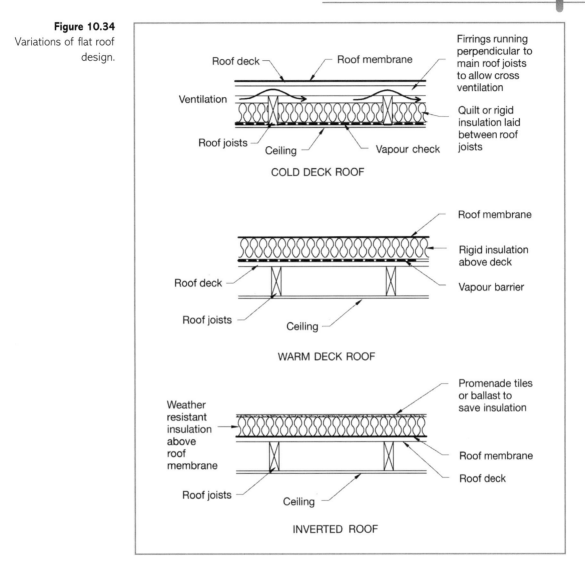

Figure 10.34
Variations of flat roof design.

COLD DECK ROOF

Roof deck — Roof membrane — Firrings running perpendicular to main roof joists to allow cross ventilation

Ventilation

Quilt or rigid insulation laid between roof joists

Roof joists — Ceiling — Vapour check

WARM DECK ROOF

Roof membrane

Rigid insulation above deck

Roof deck

Vapour barrier

Roof joists — Ceiling

INVERTED ROOF

Promenade tiles or ballast to save insulation

Weather resistant insulation above roof membrane

Roof membrane

Roof deck

Roof joists — Ceiling

inner skin of the cavity wall to provide the platform to the joists and spread their load along the length of the wall. Where joists run across the building they can be of length to create the desired roof overhang. At right angles to the joist span, short lengths of joist called 'sprockets' allow the overhang to be formed to the other two sides of the building.

The slope needed to clear rainwater from the roof (which should be a minimum of 1 in 80 gradient) is achieved by using smaller sections of timber placed on top of the main joists. These may be run either with or across the main joists and may be either tapered (firrings; Figure 10.36) or of diminishing size across the roof to create the fall.

The deck, often plywood, is placed on top of these timbers to create a level surface onto which the roof covering can be placed.

Figure 10.35
A joist-based flat roof structure.

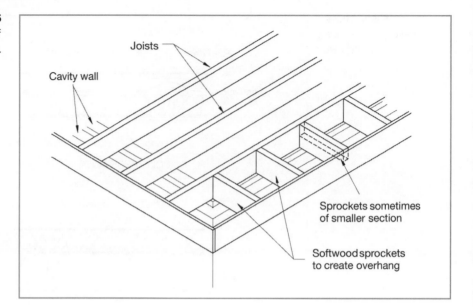

Figure 10.36
Timber firrings to provide slope.

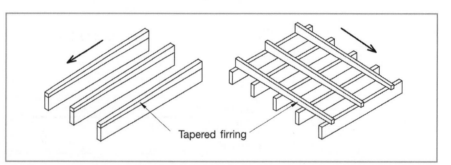

When the plasterboard is attached to the underside of the joists to form the ceiling there is a void between the joists and between the plasterboard and the deck, and it is in this space that the thermal insulation is placed. Most commonly, this is fibre quilt unrolled onto the plasterboard between the joists. If cross-ventilation through the eaves to prevent condensation is to be achieved, there needs to be sufficient space between the top of the insulation and the underside of the deck.

Other forms of flat roof structure and associated coverings are covered in *Construction Technology 2*.

REVIEW TASKS

■ Define: sprockets
            firrings
            wallplate.

■ What, apart from the deck, distinguishes a cold deck flat roof from a warm deck flat roof?

- Outline the reasons why a flat roof solution is now unlikely to be installed in a new house construction.

- Visit the companion website at www.palgrave.com/engineering/riley1 to view sample outline answers to the review tasks.

10.5 | Flat roof coverings

Introduction

- After studying this section you should be able to appreciate the basic ingredients of a built-up felt flat roof covering.
- You should also be aware of the details needed around the roof perimeter and at junctions with other parts of the building structure.

Built-up felt roofing

Bituminous felt to BS 747 is the original form of covering to flat roofs in housing and other applications. It is usual to have a number of layers in the covering, and two or three layers are typical. Within the British Standard specification there are various grades of felt, and the base, intermediate and top layers all tend to be of different grades.

The usual laying procedure is to overlap the felt as it is placed from the roll and to lay it in bituminous compound as adhesive, applied either cold or hot. As this form of roof covering may be destabilised by the heat from the sun, a layer of white stone chippings bedded in bitumen has been traditionally applied to reflect the solar heat away from the felt. The weight of the stones also tends to help hold the felt in its laid position. An alternative in more recent years has been to paint the surface of the upper felt layer with reflective silver paint.

Around the perimeter of felt roofs it is best not to position stone chippings, as exposure to weather would inevitably cause displacement, so in these areas it is normal to use a strong felt layer suitable for laying without chippings. This is miner-alised felt and easily recognisable from other felts in that it has a fine green layer of aggregate on its surface. When draining rainwater from the surface of the felt with the slopes created in the timber roof structure, only one edge of the roof will have a gutter (Figure 10.37).

The other three roof edges are generally raised in an attempt to keep the water on the roof as it moves to the gutter position. Figure 10.38 shows the timber fillet which is typically used to raise the roof edge and the sprockets needed to carry the roof structure over the external wall.

If a felted flat roof meets an abutting wall, the roof covering will be taken up the face of the wall and a flashing tucked into the brick wall will be used to prevent rain

Figure 10.37
Flat roof – eaves detail.

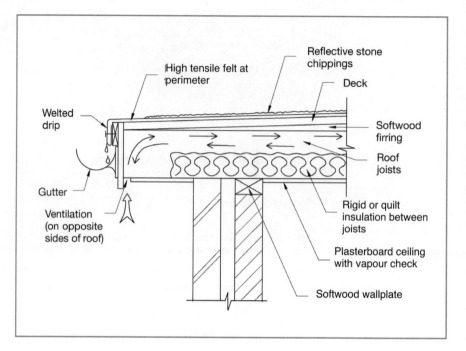

Figure 10.38
Verge detail.

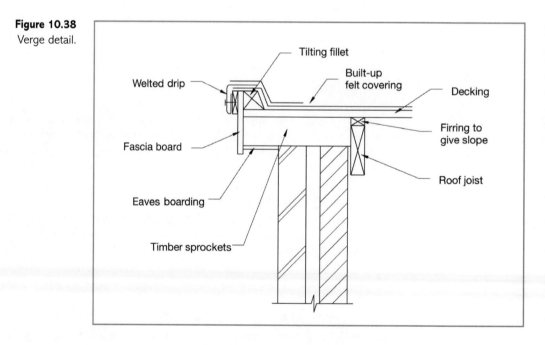

Figure 10.39
Junction between flat roof
and cavity wall.

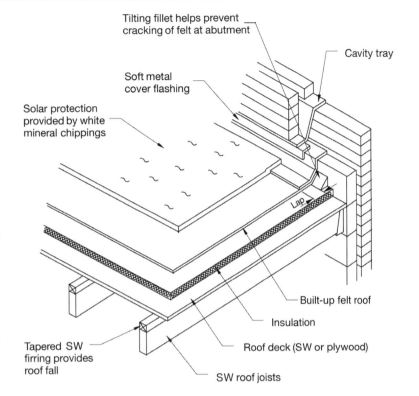

Figure 10.39
Junction between flat roof and cavity wall.

Tilting fillet helps prevent
cracking of felt at abutment

Cavity tray

Soft metal
cover flashing

Solar protection
provided by white
mineral chippings

Lap

Built-up felt roof

Insulation

Roof deck (SW or plywood)

Tapered SW
firring provides
roof fall

SW roof joists

entry. To avoid 90 degree angles in the felt, a triangular timber tilting fillet is used to break the angle, as shown in Figure 10.39.

Other forms of flat roof coverings, including some of the more modern alternatives to bituminous felt, are covered in the roofs section of *Construction Technology 2*.

REVIEW TASKS

- What do we use when roofing felt has to change 90 degrees in direction (for example, where it is dressed up against a wall)?

- Sketch a welted drip to a bituminous felt roof covering.

- Describe the benefits of high tensile roofing felts and single layer membranes when compared to traditional roofing felts.

- Visit the companion website at www.palgrave.com/engineering/riley1 to view sample outline answers to the review tasks.

PART 3

COMPARATIVE STUDY: ROOFS

Option	Advantages	Disadvantages	When to use
Pitched roof structure			
Cut roof: formed on-site	– Flexible in form – Can cope with difficult shapes – Cheap for small-scale jobs	– Expensive for larger jobs – Slow to construct – Potential for quality problems owing to site construction	– One-off or small-scale roof construction – Complex roof formations or alteration of standard trussed roof shapes
Trussed rafter roof	– Cheap for larger jobs – Roof designed by manufacturer – Loads transferred only to outside walls – Factory assembly ensures quality control – Long spans possible	– Lead-in time required for manufacture – Standard components may limit flexibility – Transport, storage and assembly of large components – Potential loss of roof void space unless using specific trusses	– Trussed roofs have become almost standard in modern house building
Pitched roof covering			
Slates	– Aesthetically pleasing – Available in a range of sizes – Easily trimmed on-site	– Expensive – Potential for damage to fixing holes – Double lap fixing slows process	– The selection of slates or tiles is a subjective decision based on aesthetics, functional requirements and cost constraints. There will inevitably be several potential options in any given case
Plain tiles	– Available in a range of colours and textures – Robust	– Many tiles required to cover a given area – Double lap fixing slows process	
Interlocking tiles	– Cheap – Single lap fixing ensures speed of fixing – Available in a range of colours and textures – Robust	– Heavy – Difficult to detail small areas	

COMPARATIVE STUDY: ROOFS (CONTINUED)

Option	Advantages	Disadvantages	When to use
Flat roof structure			
Cold deck roof	– Cheap – Easy to construct – Relatively shallow depth as insulation is contained within structure	– Ventilation required to prevent condensation – Potential for differential movement between covering and structure	– Unusual in house building – Often used for outbuildings and garages
Warm deck roof	– Does not require ventilation – Warm deck resists differential movement of covering	– Costly – Potential for traffic to damage insulation	– A wiser choice than cold deck options, but still unusual in house building
Inverted roof	– Protection of impervious membrane – No need to ventilate – Insulation is upgradable	– Difficult to detect leaks in the event of failure – Potential for damage to insulation – Requires protective surface covering – Costly	– A wiser choice than cold deck options, but still unusual in house building
Flat roof covering			
Roofing felt	– Cheap – Familiar technology	– Older forms suffer limited lifespan – Multi-layer process subject to quality control problems – Solar degradation – Problems with differential movement	– Flat roofs are rarely used in house building. They would normally only be considered for garages, outbuildings and so on
Elastomeric covering	– Often single-layer technology – Extended lifespan – Able to cope with differential movement	– Costly – Single-layer covering may suffer from localised damage	
Asphalt	– Liquid application is flexible – Monolithic application results in absence of joints	– Potential to creep at upstands – Solar degradation	– Usually concrete roof structures

PART 3

10.6 | Roof drainage and roof chimneys

Introduction

- After studying this section you should be able to identify the various details that may be employed to collect rainwater from roofs.
- You should appreciate the distances that protected fire flue enclosures project beyond the roof of dwellings.

Drainage of pitched roofs

This will include the use of rainwater gutters, downpipes, hoppers, roof outlets and special details employed with parapet roof solutions.

The traditional means of rainwater collection from a pitched roof is the rainwater gutter and downpipe system, collectively known as the property *rainwater goods*.

The gutter has been constructed out of many materials over the years – timber, lead-lined timber, asbestos cement and cast iron – but today plastic predominates. There are many different gutter profiles which can achieve quite different visual effects, but it should be remembered that the prime considerations are in relation to the carrying capacity of the gutter and its ability to handle the run-off from the roof area in question.

Figure 10.40 shows the various gutter profiles, of which the elliptical or deep flow has the biggest capacity.

Where an angular gutter profile is used, such as the trapezoidal type, the downpipe tends also to be edged rather than round. Typical plastic sectional profiles include square, rectangular and round.

Figure 10.41 shows a trapezoidal gutter with a square downpipe to illustrate some of the terminology that applies to rainwater goods.

Figure 10.40
Rainwater gutter profiles.

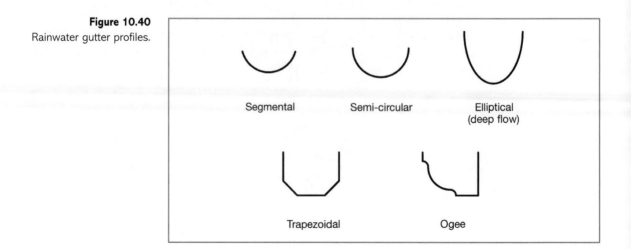

Segmental   Semi-circular   Elliptical (deep flow)

Trapezoidal   Ogee

Figure 10.41
Rainwater goods
terminology.

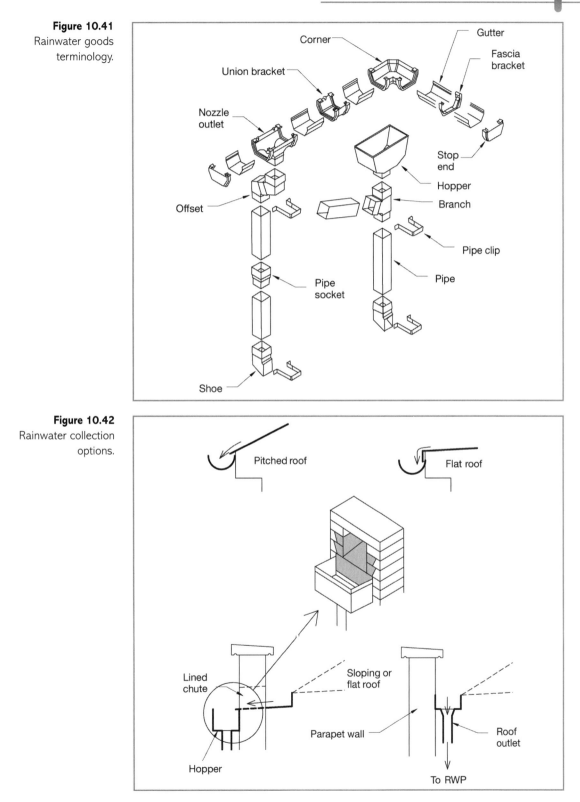

Figure 10.42
Rainwater collection
options.

In situations where the external wall of the property extends beyond eaves level to form a parapet, a gutter will be formed in the roof structure behind the parapet and connection will be made from this gutter to the rainwater downpipe. Figure 10.42 shows this detail and also summarises the collection options at the roof perimeter and behind parapets.

Rainwater collected in a parapet gutter tends to be moved to the downpipe in one of two ways. It will be taken via a chute formed through the parapet wall and into a box hopper head on the top of the rainwater pipe or it will discharge through a roof outlet accessory into an internally located downpipe.

Rainwater harvesting and recycling

It has long been recognised that the way in which water is treated and used in dwellings results in high levels of consumption of potable water. This is despite the fact that the vast majority of the water used in dwellings does not need to be potable for the use to which it is put. The practicality of providing two water supplies, one potable and one not, means that it is simply not feasible to provide mains water in other ways. However, there is huge potential for the harvesting and storage of rainwater to cater for non-potable domestic usage. The potential for such harvesting to reduce the demand for potable mains water results in tangible benefits in terms of sustainability and, since the advent of metered water supplies, cost to the occupiers. Although rainwater harvesting systems are not yet common, they are increasing in number.

The basic principle of such systems is that rainwater is collected from roofs and hardstanding run-off, and is stored and/or redistributed for non-potable use. The collection of water from roofs is more common since water from hardstanding areas

Figure 10.43
Typical rainwater harvesting system.

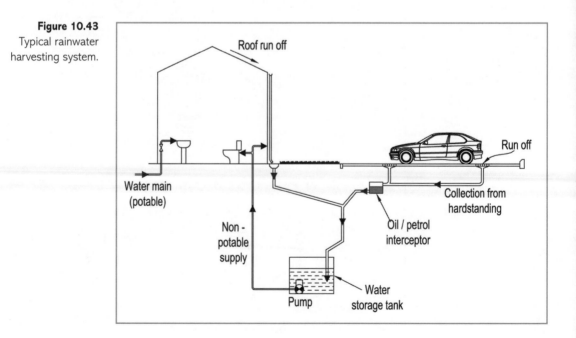

can contain contaminants that are undesirable, even in grey water. Figure 10.43 illustrates the broad principles of such systems. It is absolutely essential that the harvested rainwater does not come into contact with the potable supply.

Roof chimneys

Many modern houses now contain gas flue block systems for conveying the discharges from a gas fire located in the main lounge through to the open air at roof ridge level. Such flues are embedded in the blockwork of internal partitions as they are of similar thickness to the blocks themselves. A special ridge terminal (similar to a ridge tile) is used at the point of discharge into the air.

These flues are of limited capacity and only capable of dealing with the combustion waste products of gas appliances, which tend to be small capacity discharges. Additionally, the chance of a fire within these flues is very limited, but this is not the case for more traditional open fire flues, which will require roof chimneys to be constructed.

Open fires located in the main lounge have tended to burn coal, coke and logs, generating considerable amounts of combustion waste and needing much larger capacity flues than those required for gas appliances.

Open fire flues are lined with clay flue liners as they move towards the chimney pot point of discharge. These liners have a number of functions, but tend to contain the smoke of the combustion products. The potential for fire within these flues is quite high, and therefore the tendency is to surround the liners with an incombustible jacket of brickwork. This jacket will extend through the property, through the roof and beyond, as is needed to satisfy the Building Regulations for fire safety. Figure 10.44 gives some dimensions of the enclosing incombustible jacket for different roof situations. Remember that it is the incombustible jacket to which the dimensions relate, not the chimney pot.

Figure 10.44
Incombustible surrounds
to open fire flues.

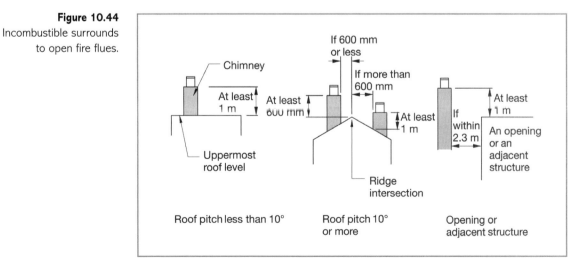

PART 3

Chimneys and flues are generally classified under three basic designations in accordance with BS EN 1443: Chimneys, general requirements. These designations are:

■ Custom built chimneys
■ System chimneys
■ Connecting flue pipes.

Typical examples of these forms are illustrated in Figure 10.45. The generic features of each of the three designated forms can be summarised as follows:

Custom built chimneys

These tend to be constructed using brick or block with a flue or liner incorporated. The space between the liner and the brickwork/blockwork is normally filled with a lightweight insulative cement or concrete.

Figure 10.45
Examples of generic chimney and flue forms as classified in BS EN 1443.

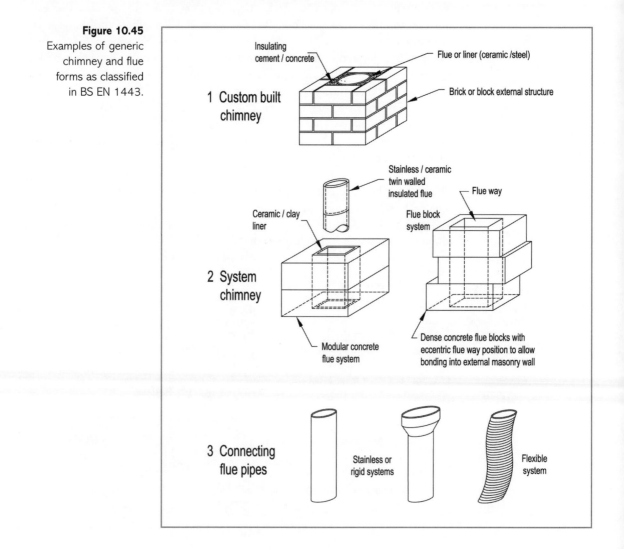

System chimneys

These may be formed using prefabricated block systems, which comprise an inner liner of ceramic or concrete material surrounded by lightweight insulating concrete. They may also include traditional concrete line blocks that connect to create a narrow flue within the wall of the property. Alternatively, they make take the forms of lined or double-walled stainless steel or ceramic/concrete flues or chimney systems.

Connecting flue pipes

These are used to connect appliances to one of the chimney/flue forms described above and may be flexible or rigid in form.

REVIEW TASKS

■ What is the collective name for rainwater gutters and downpipes?

■ Name *two* fittings used on a gutter and *two* used on a downpipe.

■ Describe where you would use a roof outlet fitting.

■ Visit the companion website at www.palgrave.com/engineering/riley1 to view sample outline answers to the review tasks.

10.7 | Thermal insulation and condensation prevention

Introduction

■ After studying this section you should appreciate the features that contribute to condensation effects.

■ You should be aware of the particular difficulties associated with condensation in roofs.

■ You should appreciate the difference between surface and interstitial condensation.

Overview – the mechanism of condensation

Dew point is the temperature at which saturated air releases water vapour as liquid. This will vary depending on individual circumstances.

Condensation arises when moisture vapour in the air is released as liquid droplets, often onto surfaces of the building interior that are cold. This release occurs at a critical temperature termed the **dew point**, which depends on the temperature and relative humidity (the amount of moisture in the air). To a point it could be said that the warmer the air, the greater its ability to carry moisture, but there is a limit which when reached would allow the air to be described as saturated.

Figure 10.46
Interstitial condensation
within a wall.

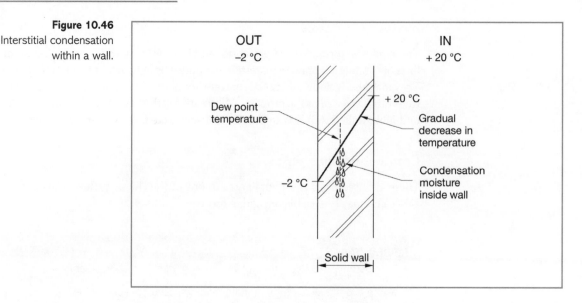

In the UK the relative humidity level is often between 50 per cent and 70 per cent, compared with locations closer to the Equator, where levels of humidity close to 100 per cent saturation are commonplace.

There are two types of condensation that we can experience: *surface condensation*, which we experience on the internal surface of glass on windows in winter, and *interstitial condensation*, which arises when the critical dew point temperature is reached in the body of the element (within the roof or within the wall).

To explain interstitial condensation it is easiest to use the situation of an external wall in winter. Figure 10.46 traces the temperature drop as we move towards the outside of the wall. For the prevailing relative humidity it could be that the *dew point temperature* occurs part way through the brickwork, and it will be at this point where moisture is released from the air, creating physical moisture and dampness within the wall.

Good levels of thermal insulation in external building elements (principally roof and external walls) will help to keep the interior surfaces of those elements above the dew point temperature, preventing surface condensation from happening and discouraging interstitial condensation in regions of the element which are close to the building interior.

Key factors in condensation control in houses are:

- Adequate insulation
- Adequate ventilation
- Adequate internal temperature
- Adequate control of generated water vapour.

We will examine each of these in turn.

Insulation and condensation

It is vital to have good levels of thermal insulation in the roof element to prevent heat loss. This will not only reduce the cost of energy but also have a beneficial environmental effect. A key role for thermal insulation to play is in helping to preserve heat internally, thereby maintaining the level of temperature on the inside of external elements such as roofs and walls. If these internal surfaces are allowed to drop to the dew point temperature, surface condensation will occur and dampness will result, causing mould. The U (thermal transmision coefficient) value for the roof advised within the context of the TER by the Building Regulations is currently 0.20 W/m^2 K, almost one-quarter of a watt of heat energy flowing through each square metre of the roof element for each degree in temperature difference between the inside and outside of the roof. One of the ways of satisfying this requirement is to supply fibre quilt insulation between the ceiling joists of the roof inside the roofspace.

Ventilation and condensation

Ventilation and air changes within the property or within an element such as the roof are vital to carry away moisture vapour that is held by the air. If this moist air can be carried to outside the building then the chance of condensation is reduced. We have to recognise that warm air rises and that as a consequence heat loss through the roof is expected to be greater than through the external walls. This is reflected in the better insulation requirements of the Building Regulations for roofs when compared with walls. When air is warmed it has a greater capacity to carry moisture (up to its saturation limit) and air moving through a ceiling into the roofspace will often contain moisture in significant quantities.

Roof design can be divided into two forms in housing: pitched roofs and flat roofs. For both of these the Building Regulations require ventilation and, ideally, cross-ventilation to carry the moisture away, as discussed earlier in this chapter.

In pitched roofs with the insulation placed typically on ceiling level, the roofspace is relatively cool in winter months while the underside of the roof covering may be very cold. Any airborne moisture finding its way to meet the underside of the roof covering is likely to condense on the inner surface of the covering and form water. By cross-ventilating via opposing eaves, air circulation can be achieved and the possibility of condensation reduced. Instead of ventilating via the eaves we may use vent tiles across the roof area to allow the moisture-laden air to escape. Figure 10.47 summarises some cross-ventilation options for different roof shapes.

Occasionally, the interior of a house may be subjected to applied air pressure by the installation of a fan (often at upper floor ceiling level) to force air into the house interior. If a positive pressure results, air will be encouraged to escape to the outside via whatever route is available (for instance, via spaces around windows and doors). These systems are called *positive pressure ventilation systems* and are generally associated with existing property which is displaying internal condensation problems.

Figure 10.47
Ventilation of different
roof shapes.

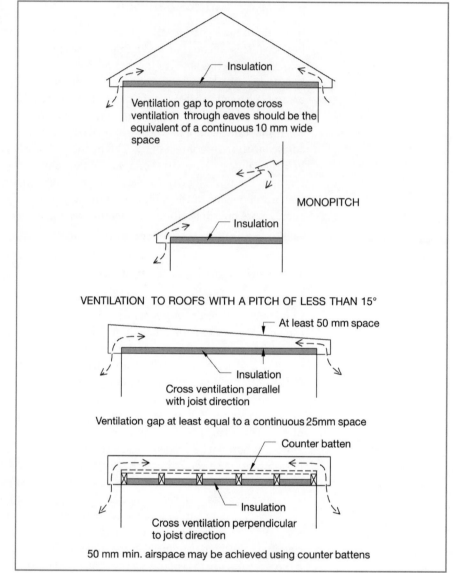

Internal temperature and condensation

One of the main issues here is the need to keep the external shell of the building above the dew point temperature. If this can be achieved then condensation will not occur.

We have to consider the form of space heating which is to be used, the nature and adequacy of the heat delivery to the rooms, and the economy of the system employed. Economy may be an issue, as we may have systems which are technically capable of maintaining the desired internal temperatures but are not fully deployed owing to the expense of the running costs. Many electricity-based space heating systems have fallen into this classification over the years, and these have given rise to the expression *fuel poverty* – a situation where the homeowner cannot afford to run the system as a result of high operating costs.

Even in situations where a desirable method of space heating such as gas-fired central heating is employed, if the system is not in operation enough during cold weather or the heating radiators are of insufficient size to heat the rooms then condensation problems may arise.

Control of generated water vapour

Certain activities in the home create extensive amounts of water vapour in the air which may potentially cause condensation. These activities include, for example, clothes' drying or bathing. If after using the bath one were to open the bathroom window and close the bathroom door then much of the water vapour would be transferred to the outside. If, however, one were to leave the window closed and leave the bathroom door open, then most of the vapour would move through the house to create great potential for condensation.

Dehumidifiers are a modern solution used for air moisture collection and may be quite effective, but they can also mask rather than cure a problem.

In houses which have inadequate methods of space heating, portable flueless bottle gas heaters are often used to supply localised heat. Unfortunately, it is known that this type of heater generates significant amounts of water vapour, which may add to the condensation problem.

The way in which dwellings are ventilated is fundamental to the control of internal moisture levels. Alternative approaches to ventilation are considered in Chapter 11.

REVIEW TASKS
- Why is it important to carry the brickwork of the chimney stack beyond the roof covering?

- What thickness of quilt insulation would satisfy a 0.20 *U* value for a domestic roof?

- Visit the companion website at www.palgrave.com/engineering/riley1 to view sample outline answers to the review tasks.

PART 3

11

Windows, doors and ventilation

AIMS

After studying this chapter you should be able to:

- Appreciate the various alternatives available for the provision of windows and doors to dwellings
- Understand their functional requirements and the limitations of the various options
- Understand the ways in which the design of windows and doors is linked to human physical attributes and limitations
- Realise the implications of adopting certain design alternatives in terms of moderation of the internal environment
- Describe the various materials and arrangements used for glazing in windows and doors
- Appreciate the need for ventilation in dwellings and the mechanisms through which it can be provided

This chapter contains the following sections:

11.1 Ventilation and dwellings
11.2 Functional performance of windows
11.3 Window options
11.4 Orientation and glazing
11.5 Door types

INFO POINT

- Building Regulations Approved Document F (2004 including 2011 amendments)
- Building Regulations, Conservation of fuel and power, Approved Document L1A – Work in new dwellings (2010)
- BS 459: Specification for matchboarded wooden door leaves for external use (1988)

- BS 644: Wood windows (1989)
- BS 1186: Timber for workmanship and joinery (1988)
- BS 1455: Specification for plywood manufactured fom tropical hardwoods (1992)
- BS 4787: Internal and external wood doorsets, door leaves and frames. Specification for dimensional requirements (1980)
- BS 4873: Specification for aluminium alloy windows (2009)
- BS 5250: Control of condensation in buildings (2002)
- BS 5278: Doors. Measurement of dimensions and of defects of squareness of door leaves (1976)
- BS 6262: Glazing for buildings (2005)
- BS 6375: Performance of windows (2009)
- BS 6510: Specification for steel windows, sills, window boards and doors (2010)
- BS 7412: Plastic windows from unplasticised polyvinyl chloride (2007)
- BS EN 14351: Windows and doors – product standard, performance characteristics (2006)
- CP 153: Windows and roof lights (1972)
- DD 171: Specifying performance of doors (1987)
- *Energy Efficient Ventilation in Dwellings– a guide for specifiers* (2006), Energy Saving Trust

11.1 | Ventilation and dwellings

Introduction

- After studying this section you should have developed an understanding of the reasons for providing effective ventilation to dwellings.
- You should appreciate the potential conflicts between efficient ventilation and energy efficiency.
- You should be familiar with some of the methods used to provide controlled ventilation in dwellings.
- You should appreciate the link between ventilation and occupant wellbeing.

Overview

Effective ventilation is essential for the provision of a healthy, comfortable internal environment within dwellings. However, there is a tension between the current drive for sustainable, energy-efficient occupation of buildings and the provision of ventilation. The basic mechanism of ventilation is the replacement of air within the building with air from the exterior. This results in the inevitable consequence that heat is removed from the interior and cold air is introduced into the enclosure, which then requires heating. It is estimated that, in the case of older buildings, as much as 20 per cent of the energy used to heat the building is lost through ventilation. The construction of dwellings tends to be relatively simple in terms of technology and there has, traditionally, been a reliance on natural ventilation. This has changed somewhat in recent times with the advent of design approaches such as PassivHaus which adopt controlled, mechanical ventilation. Natural ventilation can present an

PART 3

effective, sustainable solution for dwellings. However, there is a difficult balance that must be achieved between energy efficiency, minimum requirements for ventilation and provision of a comfortable internal environment.

It is necessary to provide effective ventilation in dwellings for the following reasons:

- Removal or dilution of polluted or contaminated air. For example, carbon dioxide will build up in the air within a dwelling and must be replaced with fresh air from the exterior.
- Provision of 'combustion air' for fires and heating appliances (this is a requirement of the Building Regulations).
- Control of moisture levels in the air. Moisture vapour is generated by cooking, breathing and other activities and, without effective removal, can result in condensation and mould growth on surfaces etc.

In order to achieve these functions, minimum levels of ventilation are required. The level of ventilation is generally defined by 'air changes per hour' – that is, the number of times that the air volume within a space is changed in one hour – or as a rate of air change in litres per second. In a dwelling, the rate of 'whole house' ventilation is generally adequate at between 0.5 and 1.5 air changes per hour.

Traditionally, the natural air movement that occurred through loose-fitting windows and doors, open flues and poor sealing of the fabric allowed for more than adequate levels of ventilation. However, as air permeability has been reduced as part of an increased energy-efficiency agenda, this is no longer the case. As such, there may be a risk in modern homes of there being insufficient ventilation. For this reason, ventilation design has now become a routine part of the design and construction of dwellings.

Providing effective ventilation

There are two broad processes by which natural ventilation and air movement occur within buildings, these are:

- Pressure (or wind)
- Thermoconvection (or 'stack effect').

The first of these occurs when the effect of wind or air pressure on the building causes air to enter through one side of the enclosure and to pass through the building. This is sometimes referred to as 'cross-ventilation'.

The second occurs as a result of temperature differences in the air outside and inside the dwelling. In this process, the effect of heating the air within the dwelling causes it to rise and escape from the upper part of the building. The resultant movement of air promotes the entry of colder air from the exterior onto the lower parts of the building.

In addition to these natural mechanisms there are several approaches based on mechanical ventilation. These often involve targeted use of mechanical ventilation in high risk areas and may involve heat recovery systems.

The Energy Saving Trust recommends a three-pronged ventilation strategy for dwellings which reflects the basic requirements of Part F of the Building Regulations as follows:

■ Targeted use of extract ventilation in 'wet' rooms, such as kitchens, bathrooms, toilets etc. These areas are where the majority of water vapour and contaminants are released into the air. The use of direct extraction to the exterior of the building allows their removal and minimises the risk of them spreading to the other parts of the dwelling.
■ Effective 'whole house' ventilation to provide sufficient fresh air from the exterior to achieve an appropriate rate of air change. This needs to be sufficient to dilute and/or remove contaminants (such as carbon dioxide) and water vapour from the internal air and to allow appropriate levels of internal air quality.
■ The use of 'purge' ventilation to assist in dealing with high levels of vapour or contaminants that may occur from time to time – for example, the removal of smoke from a kitchen after burning the toast! This is generally facilitated by allowing for opening windows etc. to 'purge' the internal air.

The specific requirements for ventilation in new dwellings are set out in Part F of the Building Regulations. This document sets parameters for ventilation and for air permeability. Typical minimum values for extract ventilation are between 6 litres/second for toilet areas and up around 60 litres/second for kitchen areas. The minimum specified rates for 'whole house' ventilation depend upon the size of the dwelling and the number of bedrooms. The minimum requirement for a two-bedroom house would be 17 litres/second, rising to 29 litres/second for a five-bedroom house.

The general rule for 'purge' ventilation is that there must be appropriate provision in each habitable room, capable of delivering at least four air changes per hour.

You will remember that the issues surrounding energy efficiency and air permeability were dealt with in Chapter 10. It is not appropriate to revisit that topic in detail here. However, it is worth reminding ourselves that there is a basic tension between providing sufficient ventilation to achieve the requirements outlined in this section and ensuring the minimisation of heat loss through uncontrolled escape of air.

Alternative approaches to ventilation

There are numerous ways to deliver appropriate ventilation levels in buildings and the selection of the most suitable option depends on many factors. The form of the building fabric, the behavior of the occupiers, the local climatic conditions and various other factors will all play a part in the final choice of design solution. There are many variations available, however they can be broadly categorised within one of the following types:

■ Background ventilation with intermittent use of mechanical extraction. This system uses 'trickle' or background ventilators in window units to allow controlled levels of background ventilation, supported by the use of mechanical extraction when needed.
■ Passive stack ventilation (PSV). This system relies on the use of a series of ducts connecting internal spaces to roof or ridge outlets that encourage air flow due to

PART 3

the combination of stack effect (thermoconvection) and wind passing over the outlets to generate a flue effect that draws air out of the building.

■ Continuous mechanical extraction.
■ Continuous mechanical supply and extraction with heat recovery.

These systems are illustrated in Figure 11.1.

Figure 11.1
Ventilation options for dwelling.

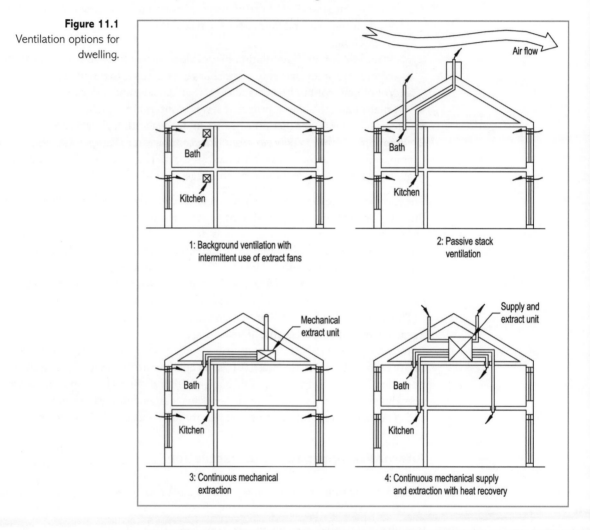

1: Background ventilation with intermittent use of extract fans

2: Passive stack ventilation

3: Continuous mechanical extraction

4: Continuous mechanical supply and extraction with heat recovery

11.2 | Functional performance of windows

Introduction

■ After studying this section you should be able to describe the ventilation and insulation requirements and limitations of windows.
■ You should be able to identify the general performance requirements of windows.

■ You should understand the implication of window design on natural lighting of building interiors.

Overview

Windows provide for many elements of the performance of the enclosure or envelope of the building. In the earlier sections of this text dealing with the general performance requirements of the building fabric, several aspects were specifically catered for by the provision of windows in the external walls, or possibly roofs. As such, the performance requirements of windows may be considered within two broad groups: performance requirements of the building fabric that are catered for by provision of windows, and specific performance requirements of windows as individual components. These may be summarised as follows:

■ Ventilation
■ Natural lighting
■ Exclusion of noise
■ Exclusion of contaminants
■ Thermal insulation
■ Security
■ Durability
■ Aesthetics/decorative application
■ Structural stability/performance
■ Sustainable manufacture
■ Buildability.

The provision of windows is essential to allow light entry and ventilation to the interior of the building, but a window also has many other desirable performance characteristics as outlined above. *Fenestration* is the term used to refer to the size, shape and configuration of windows in a building elevation. It is derived from the French word '*fenêtre*' – window.

For compliance with the Building Regulations, Part F for habitable rooms and sanitary accommodation, the requirement is an openable area of window which is equivalent to 1/20th of the floor area of the room served. It is also necessary for dwellings to be provided with windows that are of sufficient size to allow for means of escape from the upper floor in the event of a fire.

Trickle ventilation is now a routine feature of high-performance timber, aluminium and PVC window frame designs, and this may considerably help compliance with the background ventilation requirements. As windows have become more accurate in their manufacture and the use of sealing gaskets has become commonplace, the levels of background ventilation through air leakage have been greatly reduced. Hence it has been necessary to introduce controlled levels of background ventilation to ensure air movement, dilution of contaminants and reduction in condensation risk. The approach taken in dealing with ventilation and air permeability has changed greatly in recent years and a series of traditional 'rule of thumb' mechanisms associated with windows and door design has now been consigned to history. A far more accurate, calculated approach is now taken.

The positioning and configuration of windows within the building enclosure are important in relation to the overall performance of the building enclosure. The historical development of dwellings has seen considerable change in the shape and form of windows. This is, in part, associated with aesthetics but is largely the consequence of advances in the technology of lighting and ventilating buildings. In buildings that pre-date the use of electric lighting, the windows will generally be tall and narrow: tall to allow the maximum depth of light penetration into the rooms; narrow because of limited lintel technology restricting spans. As artificial lighting has evolved and lintels have allowed greater spans, windows have become wider and shorter.

When we examine light entry to the building, a feature of the window framing that applies is shaping to maximise light entry. If we were to take timber as a material to illustrate this point, the tendency is to taper the frame components rather than to leave them square edged (Figure 11.2). As can be seen, the square-edged frame would allow significantly less light into the building.

Windows are also the subject of focus for the Building Regulations, Part L (Conservation of fuel and power) in respect of heat loss, and this is primarily due to the materials used for the glazing rather than the frame material itself. If we examine the heat insulation to be provided by external elements such as the walls and roof, the insulation required is often reflected by the U value needs of the Regulations.

Approximately twelve times the amount of heat flows through single glazing compared to a modern house wall.

If we contrast the U value currently to be provided by these external elements with that of glazing, we can see that the glass allows a lot of heat to escape. The U value for single glazing, for example, is approximately 5.7 W/m^2 K. This means that glass loses around twelve times the amount of heat that is allowed to flow through the wall. Because of this, as would be expected, the use of single-glazed windows is now very rare and the impact upon the energy performance of the dwelling would be significant when using the 'whole house' method of calculation. Double glazing is now almost ubiquitous in new house construction, although the development of triple glazing has also now arisen as a result of higher levels of environmental control required by the Building Regulations to achieve the appropriate Target Emission Rates.

Resistance to weather penetration may relate to both rainwater and air entry. Timber again is a good material to illustrate some of the problems faced in these two areas, as timber is less dimensionally stable than, say, PVC sections. Timber is hygroscopic: it has the ability to absorb and lose moisture from the air and from physical contact. As the moisture content in timber changes, the dimensions of the component may change, and this is most noticeable across the grain of the wood. Such changes will of course increase or decrease the gap between the openable component and the framing in which it sits, and naturally this has to be considered when looking at penetration of rainwater and air entry under wind action. Timber opening windows incorporate throats and check throats into the framing to intercept water, and these will also play a part in resistance to air entry (see Figure 11.5).

With more modern materials, such as PVC (which by chance are also fairly stable dimensionally), the way to deal with these two issues is by inclusion of weather

Figure 11.2
Frame restrictions on light entry.

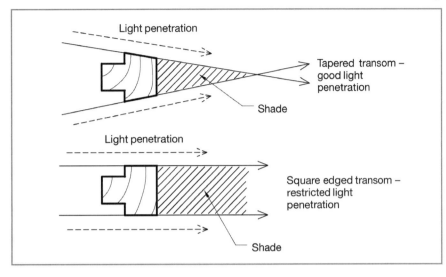

sealing between the movable components. In this way both rainwater and air are stopped at the face of the unit, and even partial penetration is prevented.

Unfortunately, such advances in performance may also create problems. A frequent example is the occurrence of condensation.

When ill-fitting timber windows are removed and tight-fitting PVC windows are installed, the natural background ventilation of the room in question can change dramatically. Although the ill-fitting windows created draughty conditions and undesirable heat loss, they can provide a useful source of air change within the room. Once this is removed, the more stagnant air conditions that may arise can be a significant contributor to condensation and mould growth. Examination of maintenance of air change rates when about to fit PVC windows is often a consideration of local authority housing improvement schemes.

We could say that one of the desired features of a window in terms of functional performance is durability, and to an extent this characteristic alone has been an influence on the way in which the market for different window types has developed. After the Second World War the predominant window material used was timber, and of the timber varieties softwood was the most widely used material. In the last twenty years or so of the previous century, hardwood (principally Brazilian mahogany) became fashionable for its appearance and its durability. It was at about this time that energy conservation became a national issue following the fuel crisis of the 1970s, and double rather than single glazing was in demand. These hardwood units required considerably less maintenance than their softwood counterparts, and oiling or staining appeared to be far more popular than painting. Matters moved on in the same direction as double-glazed units in PVCu started to emerge with the advantage of no exterior application for preservation, just washing. The increases in requirements for thermal insulation and air-tightness have continued to drive the development of these windows as they have good sealing properties and can be manufactured to high levels of accuracy. As technologies associated with window design have improved the emergence of 'high-performance' timber windows has

PART 3

become evident, and these have been successful in maintaining the market share for timber windows in the new house building sector.

Security has been an issue with windows for many years. The alternative to putty (linseed oil plus chalk) for glazing is to hold the panes of glass in position with beads. Beads may be nailed (bradded), screwed or clipped into position (PVCu).

For security purposes those windows which use beads as a pane-fixing option tend to locate the beads internally now rather than externally, as was the original practice.

Opening windows are often restricted in size or in the extent to which they open in a reflection of the need for security. Locking is now a common feature rather than the use of casement stays with pegs and casement fasteners.

REVIEW TASKS

- Compare the likely U value of single glazing with that of a modern external wall and suggest the likely relative performance.

- For habitable rooms, what fraction of the floor area is to be the equivalent openable area of window?

- Visit the companion website at www.palgrave.com/engineering/riley1 to view sample outline answers to the review tasks.

11.3 | Window options

Introduction

- After studying this section you should understand how the limitations on window size have evolved from the capabilities of the user.
- You should appreciate the differences in the sectional shape of the frame caused by the use of different window materials.
- You should have an appreciation of the different options available for window design and manufacture, and should appreciate the performance characteristics of each type.

Overview

The main distinction between different types of window has emerged from the different ways a window can be arranged to open (Figure 11.3). Additionally, there tends to be a division of types caused by the supporting mechanism for the opening component. If an opening window is supported on hinges, we tend to call those openers *casements*. By contrast, if the window opens using some other form of support (such as a pivot), then we tend to call these windows *sashes*.

The amount of window that opens outwards has to be limited for at least two reasons: human reach and possible wind action on the component as opened, which

Figure 11.3
Opening window design
options.

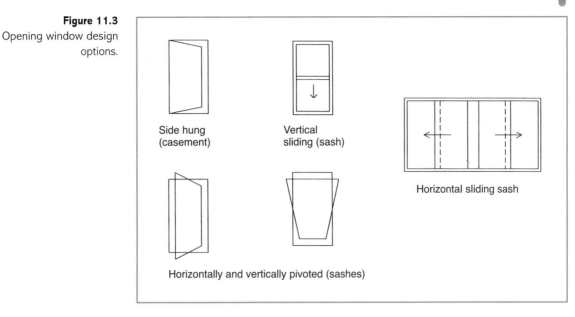

could make it very difficult to handle if large. As a general rule, this limits the amount of opening outwards to around 600 mm. Note the varying amount of opening for ventilation that we have when opening a side-hung casement against a vertically pivoted window. In the latter case, half of the window opens inwards as the other half opens outwards, creating perhaps 1,200 mm of ventilation opening.

As designs and the technology of supportive ironmongery have advanced, we now have a range of tilt and turn windows which can be moved to allow internal cleaning of the exterior face.

Timber windows

BS 644 is the standard applied at present for timber windows. These are commonly arranged into a number of fixed and opening lights.

As shown in Figure 11.4, the opening lights may be identified using broken lines, the apex of which points to the edge where the hinges are located.

Examination of the framing of a standard timber section shows, as mentioned earlier, the tapered sections of timber machined for the window in order to maximise the light entering the room from the unit. In Figure 11.5 the top hung casement is shown in section as a *vent light*. This type of sectional profile of the supportive perimeter frame and opening components is also often referred to as 'double rebated', as rebates occur on the main frame and on the casements themselves.

The traditional house window, emerging around the mid-19th century and still popular today, is the vertical sliding sash window. This vertical movement arrangement is unusually effective for a material which is known to expand and contract as it loses and gains moisture. Friction is one force which holds the window in position when opened, and this, together with the weight of the glazed sash, need to be

PART 3

Figure 11.4
Casement window terms.

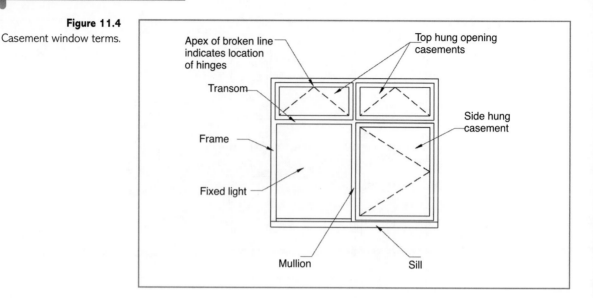

overcome by the user of the window. With this in mind, the original designers of the window included cast iron balancing weights attached to the sliding sash via a cord and pulley. The pulley and weight are located in the box side frame which was a feature of this window type. The rope cord passed over the top of the pulley and out of the box frame, attaching to the side of the rebated sash as shown in Figure 11.6.

Although the sliding sash window proved to be a considerable success as a design, broken sash cords in the traditional window are often encountered in old examples of this window style.

Figure 11.6 shows the traditional vertical sliding sash window in sectional elevation and in plan section towards the left-hand side of the diagram. Although this window with its cast iron counterbalance weights works very well, the rope sash chords used between the sliding sashes and the weights tend to deteriorate over time, and broken sash cords are common. To replace the cord is a laborious task, as we have to dismantle the box frame to allow access to the cast iron balance weight. Mindful of this, the modern equivalent sliding sash uses balance springs rather than weights and a more conventional frame rather than a box frame. The balance spring casing is attached to the side frame and the end of the spring to the underside of the sash. When the window is closed the spring is at its maximum extension, ready to help counteract the weight of the sash when in motion.

Advances in the manufacturing techniques associated with timber windows have resulted in the development of high-performance engineered timber windows. These provide the same degree of fine-tolerance manufacture and performance as PVCu or aluminium windows and perform extremely well in terms of thermal efficiency and air permeability. They also offer a sustainable design solution. Figure 11.7 illustrates some of the key features of engineered timber window units.

There are a range of alternatives to timber (which includes of course both softwood and hardwood). These include (Figure 11.8):

Figure 11.5
Casement terminology.

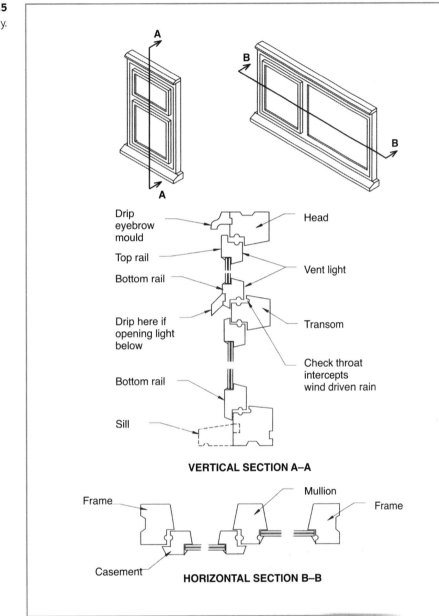

Drip
eyebrow
mould

Head

Top rail

Vent light

Bottom rail

Drip here if
opening light
below

Transom

Bottom rail

Check throat
intercepts
wind driven rain

Sill

VERTICAL SECTION A–A

Frame

Mullion

Frame

Casement

HORIZONTAL SECTION B–B

Steel

Steel windows are uncommon in modern house construction because of the high levels of heat transmission through the frame, resulting in difficulties with cold bridging. Although the use of galvanised steel frames was once common, they are now found only in older properties. Steel windows use predominantly N or Z sectional profiles, where a pivoting sash closes on its frame.

PART 3

Figure 11.6
Vertical sliding sash
options.

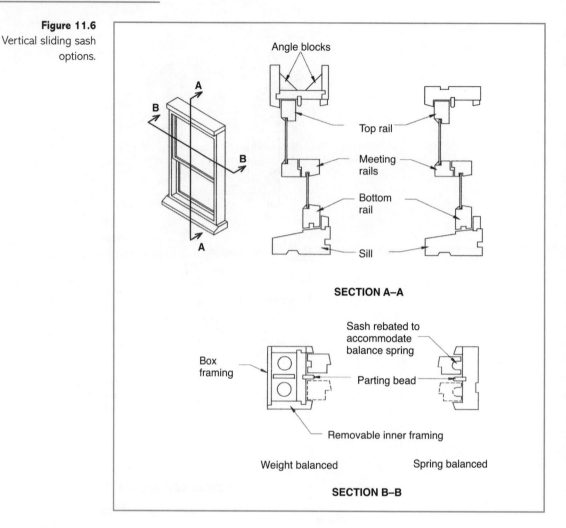

Figure 11.6 Vertical sliding sash options.

Aluminium

Aluminium double-glazed frames were once in the vanguard of double-glazing developments in house construction. The ability to manufacture robust frames with high levels of accuracy, resulting in tight sealing against draughts and the ability to create large window sections easily resulted in aluminium being a popular choice for early adopters of double glazing. The use of powder-coated aluminium frames is still very common in commercial and industrial buildings but their use in housing has diminished considerably. Aluminum window sections tend to be box sections for strength, and although the extruded section seems complex, it is still basically a box.

PVCu

PVCu windows are perhaps the most commonly used in modern house construction. The tight tolerances achieved, effective weather sealing and low thermal transmis-

Figure 11.7
Engineered timber
windows.

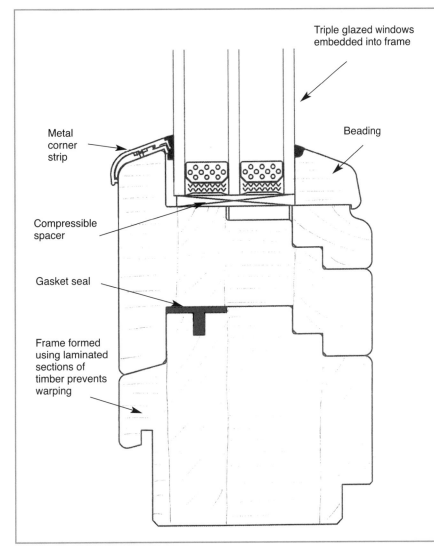

sion through the frame combine with low maintenance requirements to make this type an effective and economical window solution. PVC units tend to be box-like in section, containing steel to improve its strength capabilities, and give the performance that is necessary for security. A range of different profiles exists, although the majority are fairly similar in appearance. It is generally the case that window sections of PVCu are bulkier than equivalent aluminium or timber sections. This can have an impact on the appearance of the overall window unit and smaller casement sections can result in limited glazing areas. The relatively stable thermal dimensional performance of the windows, allied to the ability to manufacture to fine tolerance, means that PVCu units offer high levels of performance and good control of air permeability. Although high performance timber units are increasing in popularity, the use of PVCu is by far the most common in modern house building.

Figure 11.8
Frame sections – different materials.

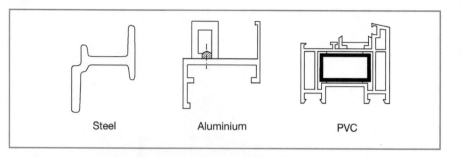

Steel   Aluminium   PVC

REVIEW TASKS

■ Why is the frame used for a window often tapered towards the inside of the property?

■ Apart from timber, name *three* other materials used for windows.

■ What distinguishes a 'casement' from a 'sash' window?

■ Visit the companion website at www.palgrave.com/engineering/riley1 to view sample outline answers to the review tasks.

11.4 | Orientation and glazing

Introduction

■ After studying this section you should appreciate the importance of orientation to the performance of windows.

■ You should be aware of the evolution that has occurred in glass production, the range of glasses that is available, and the methods of fixing the glass to the window component.

Overview – orientation

The orientation of a building is the way that it faces relative to the path of the sun. As the sun rises in the east and sets in the west, passing south as it moves, any elevation facing north will not receive any sunlight directly onto its walls. This can have a considerable influence on the amount of solar heat entering a building via the windows, depending on where the windows are located and their size.

If the main front elevation of a house faced north, there would be a tendency to place the bulk of the windows in the north and south elevations, with far less glass in the east and west elevations. Although no sunlight will be incident on the north-facing windows, considerable amounts would penetrate the windows facing south and rooms would often be purposely located facing south to gain the benefit of this sunlight.

In the study of energy conservation we must recognise useful heat gain as well as heat loss, and this has been an issue in previous years for buildings other than houses. Today, modern energy-absorbing design is starting to take advantage of sunlight entry even in housing to minimise the need for artificial heat energy, with the associated benefits for the environment.

Glass forms

Before the advent of the float glass production process, which was patented by Pilkington of St Helens, all glass was produced by the rolling process. As this did not produce a pane with parallel sides, the glass had various degrees of image distortion and a range of qualities emerged, such as SQ (selected quality) and SSQ (special selected quality). In certain locations, such as shop and store windows, visual distortion was not tolerated and the only way parallel-sided glass could be assured was by grinding and polishing flat. *Polished plate* glass was obtained in this way. Today, only patterned glass is produced by rolling.

The float process uses a molten bath of tin on which the molten ingredients of glass float, hence the name. The result is glass which is perfectly parallel on its opposing sides and of a thickness which is determined by the speed of drawing through the firing chamber across the molten tin. Thicknesses with the float process can be as little as 2 mm (agricultural) and as large as 25 mm (cladding).

Glass tends to be labelled *transparent* (one can easily see though it), *translucent* (allows light through but distorts the visual image – for example, patterned glass for use in bathroom windows and the like) and *opaque* (one cannot see through this because of the colour pigment added to the mix). In addition to these classes, we also have special glasses for solar control and for energy conservation.

The solar control range largely generates three classes of glass: *surface modified*, *body tinted* and *laminated*. With surface modified glasses, a reflective deposit is placed close to the surface of the pane to decrease the amount of solar light and solar heat allowed into the building. With body tinted glasses (usually green or smoke grey), a metal pigment is introduced to the glass mix. This tends to work on the pigment, absorbing solar heat, holding it, and then re-radiating it outwards. Good levels of light control may also be effected using this type of glass. The laminated glasses can provide the greatest degree of control on the amount of solar light and solar heat entering the building. As the name suggests, these panes are produced from two pieces of glass fused together with a metal-based deposit placed between the laminates. Colours are varied by the nature of the metal used and are typically gold and bronze. Such glass can prevent large percentages of light and heat from penetrating the building (>> 80 per cent), primarily to prevent excessive glare or heat gain, and has particular application in areas of the world with much hotter climates than does the UK.

Energy conservation glass is also surface modified to reflect internal heat back into the building. By comparison, about twelve times more heat will flow through

PART 3

single glazing, which has a U value of around 5.7 W/m² K. If we double-glaze with ordinary glass, the U value can be improved to around 2.0. Double-glazed units which have an energy control glass as the inner pane should achieve less than 2.0 grading, which represents a significant improvement to heat loss when compared with single glazing.

The influential factors in heat insulation will be the thickness of glass, the arrangement of glass (single, double or triple glazing) and the characteristics of the glass (for example, with or without reflecting additives).

In glass supply there is an inseparable link between the size of a glass pane and its thickness. As a result, when you order large panes of glass they will be of thickness predetermined by the manufacturer.

The strength characteristics of the glass may also be changed by processes subsequent to the initial formation of the pane. For example, further controlled heat treatments will make the pane more resistant to impact shocks, and such *toughened glass* would now be a standard feature of glass when used in large panes in the home for patio doors and the like.

Glazing techniques

The quality standard for glazing windows and doors is Code of Practice (CP) 153. The traditional way to glaze timber windows was with a putty seal to the frame. Figure 11.9 illustrates the putty glazing process.

Putty tends to age with exposure to the elements and eventually becomes brittle and shrinks. Once rainwater can penetrate the putty down to the timber frame rot may commence, so replacement and regular inspection are essential.

Figure 11.9
Traditional glazing with putty.

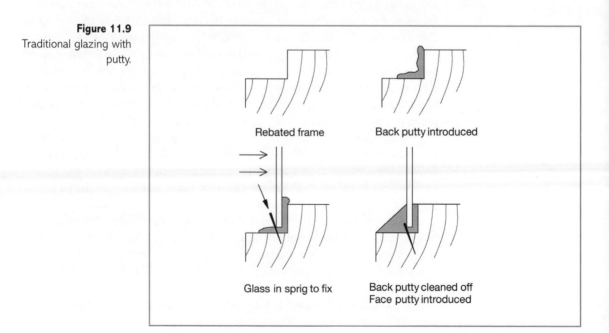

Rebated frame

Back putty introduced

Glass in sprig to fix

Back putty cleaned off
Face putty introduced

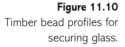

Figure 11.10
Timber bead profiles for
securing glass.

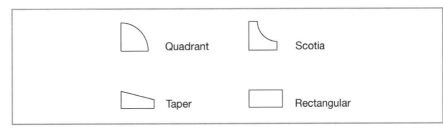

As an alternative to putty we may hold glass in position with timber beads of a variety of profiles, for example quadrant, scotia, moulded and tapered (Figure 11.10).

With external components, such as windows, the normal position for beads holding glass panels in position is on the inside of the component. This provides the best security. Of course, there will still be a need to make the pane of glass resistant to moisture entry with sealant, which currently may take the form of non-setting compounds or preformed strips. The latter tend to be U-profile in shape and are simply pushed onto the perimeter of the pane before positioning.

Beads may be fixed by nailing or screwing (for timber) or by clipping when using PVC, high-performance timber or aluminium.

REVIEW TASKS

■ Define 'orientation'.

■ Compare the *U* value characteristics of single, double and triple glazing.

■ Define 'translucent' as applied to glass.

■ Visit the companion website at www.palgrave.com/engineering/riley1 to view sample outline answers to the review tasks.

11.5 | Door types

Introduction

■ After studying this section you should be able to describe the functional performance generally required of doors.
■ You should be able to identify the range of doors typically available for domestic use and you should appreciate the difference between frames and linings.

Overview

There are a number of different types of door suitable for use in houses, and the range may be extended by considering the various materials and finishes also available.

You should appreciate typical door sizes, and the importance of the hygroscopic nature of timber in door design.

Some of the functions that doors need to fulfil are purely functional (for example, allow access), while others are non-functional (for example, aesthetics or appearance). Additionally, sometimes doors are divided into two classes before considering function, namely external doors and internal doors. Performance for an exterior door may involve more criteria than for its internal counterpart.

External doors need to:

- Provide security
- Be acceptable in appearance
- Preserve the internal heat of the building
- Maintain levels of sound insulation
- Have durability in respect of the weather
- Be of sufficient width to allow the entry of household items such as furniture as well as pedestrians.

Particularly in light of the need to be secure and to resist deterioration by weather, external doors tend to be more robust, heavier and thicker than internal doors. Typical widths include 762 mm, 838 mm and 914 mm, while height tends to be consistent at 1,981 mm, and thickness between 40 and 54 mm.

The aesthetics or appearance of a door may be varied significantly by its classification of type; panelled doors are quite different in appearance from flush doors. The material from which the door is made also reflects significantly on appearance: hardwood is quite different from PVC, for example.

Both heat insulation and sound insulation will be affected not only by the body construction of the door but by the quality of fit between the door and its frame. When timber doors are used an additional factor is the nature of the timber itself, as timber tends to be *hygroscopic* – it has the ability to absorb and lose moisture. As shown in Figure 11.11, the dimensional changes that occur as timber gains moisture and expands, and loses moisture and contracts, are not significant along the grain. However, across the grain dimension changes as large as 5 per cent could be expected.

In this situation the gap between the door and the frame may change, allowing changes in resistance to heat flow and sound penetration. Clearly, materials which are more dimensionally stable, such as PVC, have an advantage in this respect.

Durability mainly relates to resistance to the weather and to water in particular. Here again the hygroscopic nature of the material may have an effect on external doors if they are not maintained properly (with paint, varnish, preservative stain and so on).

Certain timbers, such as hardwoods, are quite resistant naturally to moisture damage. If plywood was to be used in exterior doors, a suitable grade could be chosen for its moisture-resisting qualities. Marine ply or WBP ply (water and boil proof) have high moisture resistance; see BS 1455.

When selecting doors for external and internal use, their width needs to be considered for access for furniture such as settees, sideboards and wardrobes.

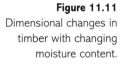

Figure 11.11
Dimensional changes in timber with changing moisture content.

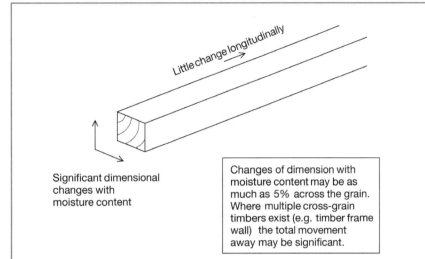

Little change longitudinally

Significant dimensional changes with moisture content

Changes of dimension with moisture content may be as much as 5% across the grain. Where multiple cross-grain timbers exist (e.g. timber frame wall) the total movement away may be significant.

REVIEW TASKS

■ Name *three* performance characteristics that you would require from an external door.

■ What is *hygroscopicity*?

■ Visit the companion website at www.palgrave.com/engineering/riley1 to view sample outline answers to the review tasks.

Panelled doors

When we consider the types of door available it is best to commence with consideration of panelled doors, as the basic member composition tends to be used in other door types, as will be illustrated shortly.

The panelled door, as shown in Figure 11.12, has a basic perimeter which consists of two vertical stiles, a top rail and a bottom rail. The stiles and top rail are typically around 100 mm in width, while the bottom rail is traditionally wider, often 200 mm or more. If the door is to have a single panel, as in the figure, the perimeter frame is usually rebated to take the panel, and the panel fixed in position using beads. The bead profiles (Figure 11.13) may vary, as may the method of fixing them in position.

If beads are nailed in position we tend to refer to them as bradded; alternatively, they may be screwed or fixed with cups and screws. The circular cups are used in conjunction with countersunk screws to allow for easy removal. This is necessary, as the head of the screw is not able to bite into the timber.

Figure 11.14 shows the creation of more than one panel using a middle rail, which traditionally is not in the middle of the door, intermediate rails (which are often little more than glazing bars in size), and a muntin used to divide the width of the

PART 3

Figure 11.12
Ingredients of a panelled door.

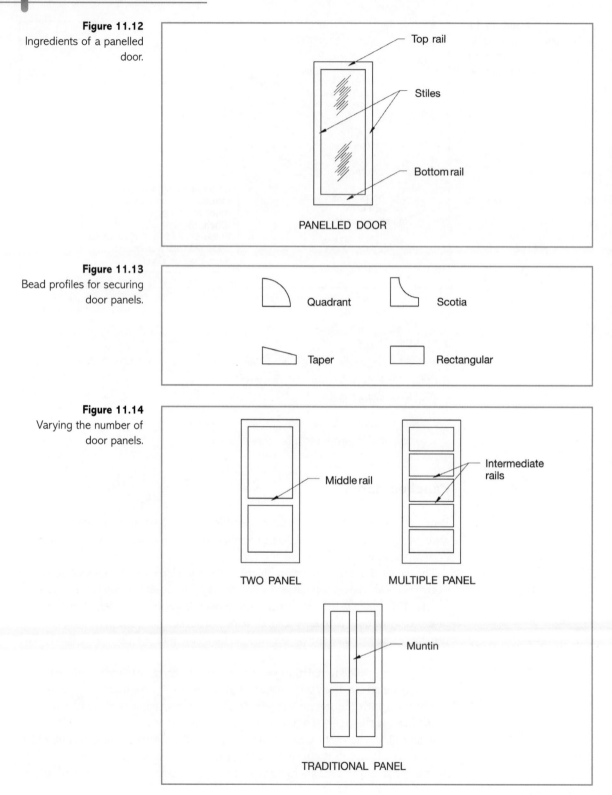

PANELLED DOOR

Figure 11.13
Bead profiles for securing door panels.

Quadrant

Scotia

Taper

Rectangular

Figure 11.14
Varying the number of door panels.

Middle rail

TWO PANEL

Intermediate rails

MULTIPLE PANEL

Muntin

TRADITIONAL PANEL

Figure 11.15
Traditional timber door
panels.

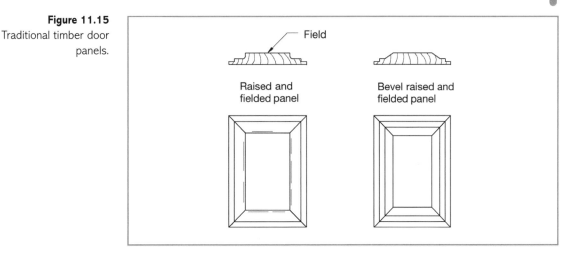

door. For two- and multiple-panel doors the bead is used extensively as the method of panel fixing, while with the four-panelled door the timber panels tend to be fixed into grooves in the surrounding members, as the door is assembled at the factory. The traditional four- (and six-) panel door will use panels which are shaped to enhance the appearance of the door and suggest strength.

'Raised and fielded' and 'bevel raised and fielded' are two of the traditional panel shapes used for four- and six-panelled doors (Figure 11.15). As shown, the field of the panel is the flat section in the panel centre. This type of traditional panel profile can be enhanced even more by adding an ornamental bead around each panel perimeter. This bead is often termed a 'bolection moulding'.

Flush doors

Most flush doors (Figure 11.16) tend to use the internal perimeter timbers that feature on the panelled door, namely stiles and top and bottom rails. The name of this door emerges from the face finish, in that it is flat.

The figure shows that there are basically four forms that the flush door takes: hollow, skeleton, cellular and laminated. Forms of hollow door tend to be the cheapest available, while the laminated core is at the other end of the cost range.

Facings to flush doors are numerous, from hardboard (the cheapest door type) to plywood, veneered plywood and plastic laminate. With the hollow door there is little support to the face covering of the door, and this makes it vulnerable to damage if covered in one of the less strong facings such as hardboard.

If this door type is to be used externally it would require a facing capable of weather resistance, and an appropriate grade of plywood would be ideal. When used as an internal door, the appearance can be enhanced considerably by the use of hardwood veneered plywood facings.

The skeleton core uses internal horizontal members to provide some support to the face covering, and the cellular core can provide a more extensive face support as

Figure 11.16
Different forms of flush
doors.

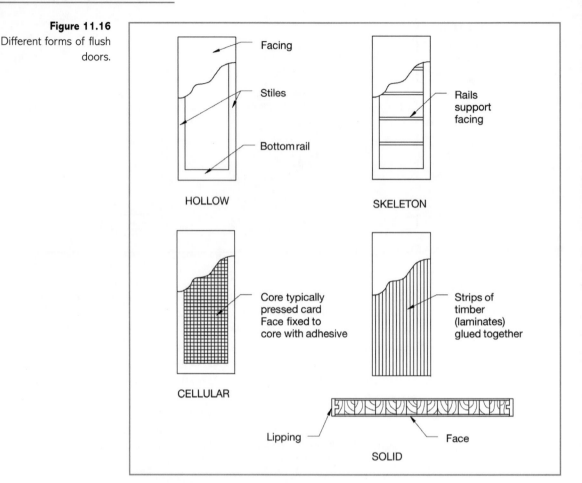

an alternative core. Different shapes of cellular core tend to be used by the different door manufacturers: diamond, honeycombed and square.

Where strength is needed from this door type, the laminated door may be used. The laminate strips of timber which are glued together to complete the core have the grain of each laminate rotated to provide strength.

To prevent wear damage to the face covering of the door, a hardwood lipping is attached to the edges (usually only the two long edges) which extends through the full thickness of the door (Figure 11.17). If the two long edges of the door are protected in this way they will resist damage as the door opens and closes to its frame.

When we provide a door with handles (lever furniture) these are attached to a metal lock assembly which is recessed into the body of the door. These tend to be called 'mortice locks' and 'mortice latches'. As the joiner places this into the door body he will form a sinking or mortice in the door edge. With panelled doors, the stile is generally not wide enough to accommodate the mortice lock or latch, and we therefore have to place an extra piece of timber inside the door during manufacture to take the depth of the mortice assembly. This required lockblock is also shown in Figure 11.17.

Figure 11.17
Lipping edges of flush
doors.

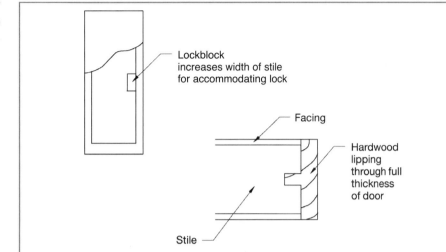

Figure 11.17
Lipping edges of flush
doors.

Matchboarded doors

Matchboarded doors (Figure 11.18) were formed originally out of floorboarding and used for exterior access doors to outbuildings and the like. The simplest form of matchboarded door possible is the ledged door, where just two cross-ledges are used as the means of holding the boarding together.

Figure 11.18
Forms of matchboarded
door.

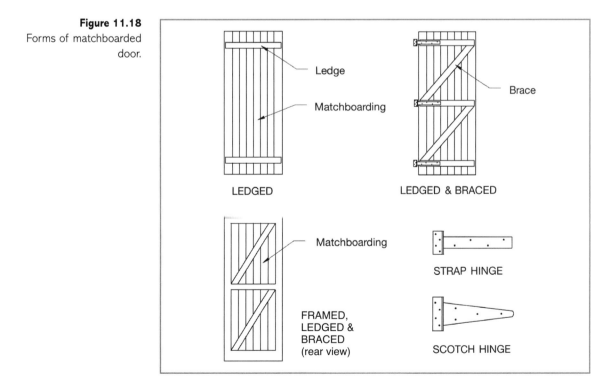

PART 3

For greater stability and strength, braces were added as shown in the figure. Note the position of the hinges, with the braces sloping up from left to right. The ledge is the typical fixture point for the hinge. The best overall appearance and strength are achieved with the framed, ledged and braced door, where effectively a perimeter frame is fitted around the matchboarding. From the front, the door has a fairly flush face, with the matchboarding finishing in line with the perimeter-enclosing framing. It is often the case that the matchboarding is tongued and grooved (just like floor boarding) and V-jointed to enhance the appearance of the joint between boards.

REVIEW TASKS

- Name *two* cores that could be used to flush doors.

- What is *lever furniture* as applied to doors?

- Visit the companion website at www.palgrave.com/engineering/riley1 to view sample outline answers to the review tasks.

Frames and linings

Doors are traditionally hung from a timber frame which is machine rebated to take the closing edge of the door (Figure 11.19).

True frames tend to be found to external doors, where the overall dimensions tend to be approximately 100×75 mm. The gap between door edge and frame is needed to accommodate changes in frame dimension that are caused by changes in the frame and door moisture content. This will ensure that the door can still be opened and that it does not have a tendency to stick (see BS 1186 for further information regarding moisture contents in timber at the time of manufacture of joinery components). Clearly, the gap is also important in relation to preserving internal heat and in relation to sound entry, and it is often therefore the case that a compressible material is applied between the frame and the door edge to assist this as *draught-proofing*. As an alternative to providing doors and frames as separate items we can order a door already hung from a frame, and this is called a *doorset*.

While frames are used to external doors, a lining is more typical to internal doors. When an opening is formed in an internal wall or partition, the timber lining will not only support the door but will also line and neatly finish off the opening.

Figure 11.19
Door frame.

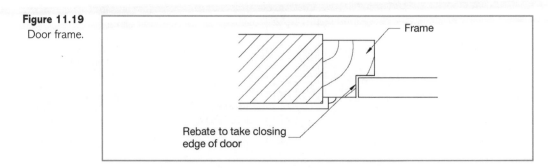

Frame

Rebate to take closing
edge of door

Figure 11.20
Door linings.

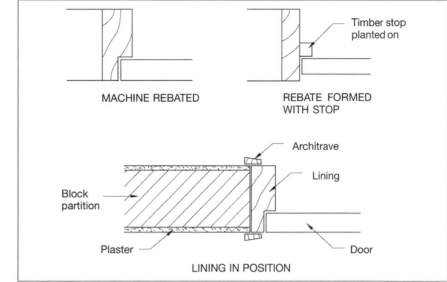

Figure 11.20 shows the use of a lining to an internal door located in a blockwork partition. Here the extreme dimension of the lining is determined by the overall width of the partition. If the blockwork is 100 mm wide with 13 mm of plaster on either side, the lining would be finished to 126 mm to extend through the full width of the opening. Where the wall plaster meets a timber material such as a lining, there will always be a crack, and to hide this junction a timber architrave is used, as illustrated.

Like a door frame, the timber lining may have a rebate to receive the closing edges of the door, which is formed by machine. Alternatively, a rebate may be created by simply fixing on (planting on) another small section of timber as a stop.

Engineered and composite doorsets

The need for higher levels of thermal performance, air permeability control and security has resulted in the development of engineered door assemblies. These doorsets comprise a frame unit, generally in PVCu with a pre-hung door that is accurately matched and sealed to the frame. Modern composite doorsets are available in a range of materials and styles but they are often based on GRP/PVCu composite panels with internal insulation and reinforcing steel. They are also provided with integrated multi-point locking assemblies.

REVIEW TASKS

■ Where would you expect to find a door frame used and where would you expect to find a door lining used?

■ For what reason would you use a *planted stop*?

■ Visit the companion website at www.palgrave.com/engineering/riley1 to view sample outline answers to the review tasks.

PART 3

COMPARATIVE STUDY: DOORS AND WINDOWS

Option	Advantages	Disadvantages	When to use
Internal doors			
Hollow core flush doors	– Cheap – Readily available in a variety of finishes – Often require no decoration	– Less durable than solid doors – Poor sound insulation – Fixings and handles must locate pre-positioned blocks	– These are the most common form of internal doors in dwellings – Used where sound and fire resistance are not issues for consideration – Almost ubiquitous in modern speculative building
Solid timber doors	– Durable – Good sound resistance – High-quality appearance – Possible 'Fire Resistance' rating if specifically designed as fire-resisting doors – Flexibility in position of fixings and ironmongery	– Expensive – Often require applied decorative/protective finish	– Selection is an issue of personal preference, although the use of solid doors tends to be more common in higher quality dwellings – Where there is a requirement for fire resistance or sound insulation, solid doors are generally utilised
Glazed doors	– Allow the passage of light to areas that have limited access to natural light	– Potential for danger since the glass is at low level. This is generally avoided by use of safety glass, as required by the Building Regulations	– Where natural light is to be maximised or when aesthetics demand
External doors			
PVCu	– Durable – No decorative maintenance requirement – Security features	– Appearance very modern – Costly – Difficult to cater for size adjustment	– Increasingly common in dwellings because of its availability and security advantages
Solid timber	– Traditional appearance – Durable, particularly if of hardwood form – Security features – Ability to adjust size on-site by trimming	– Costly if made in hardwood – Requires protective finish for durability	– Popular in a wide range of applications – Subject to personal choice
Composites	– Durable – Secure – Maintenance free – High security – Pre-hung assembly	– Costly	– Increasingly used in new dwellings

12

Internal finishes

AIMS

After studying this chapter you should be able to:

- Appreciate the range of internal finishes typically applied to residential property
- Understand some of the criteria considered before selecting finishes
- Appreciate the types of wall plaster that could be used, the range and application of plasterboard for wall and ceiling finishes, and the use of a limited range of floor finishes

This chapter contains the following sections:

INFO POINT

- BS 1187: Specification for wood blocks for floors (1959)
- BS 1191: Specification for gypsum building plasters (1973)
- BS 1230: Gypsum plasterboard. Specification for plasterboard excluding materials submitted to secondary operations (1985)
- BS 5385: Wall and floor tiling. Code of practice for the design and installation of ceramic floor tiles and mosaics (2009)
- BS 6431: Ceramic wall and floor tiles (1991)
- BS 8000: Part 11: Workmanship on building sites. Code of practice for wall and floor tiling. Ceramic tiles, terrazzo tiles and mosaics (2011)
- BS 8201: Code of practice for flooring of timber, timber products and wood based panel products (2011)
- BS 8212: Code of practice for dry lining and partitioning using gypsum plasterboard (1995)
- BS EN 13914: Design, preparation and application of external rendering and internal plastering (2005)

12.1 | Functions of finishes and selection criteria

Introduction

■ After studying this section you should have an appreciation of the range of issues that may be examined before selecting internal finishes.

Overview

The basic function of internal finishes tends to be twofold: to create internal surfaces which may be kept clean with reasonable ease, and to create internal surfaces that are visually acceptable. In dwellings, the range of finishes to certain elements such as ceilings and walls is fairly limited, but slightly more scope exists where selecting finishes for floors. The purpose of this section is to provide a very superficial overview of this topic.

Criteria

The criteria considered to satisfy ease of cleaning will tend to include:

■ Smoothness/texture
■ Absorption characteristics
■ Suitability for decoration
■ Durability.

Examination of wall finishes, for example, would involve consideration of plaster as an accepted traditional material. Plaster is smooth and tends therefore not to harbour dust. Although it is porous to moisture it can be painted to reduce its porosity if necessary. It is ideal as a smooth finish for the application of accepted decoration such as wallpaper, and despite its natural porosity it is a fairly durable material in that its surface in particular is fairly hard.

Desired criteria to satisfy the visual acceptability of the finishes would typically include:

■ Smoothness/texture, again
■ The traditional nature of the finish.

By their nature, finishes provided within property tend to be smooth, while the decoration or subsequently applied material (such as carpet) may be the ultimate and variable finish provided. To an extent, the internal finish is the base for other materials subsequently applied – materials like those specifically considered in this section.

Most internal finishes used in houses today have evolved over time as the accepted norm, and it should be recognised that the floor element often has no internal finish, simply the finish resulting from the selection of a floor solution (floorboarding, or perhaps power-floated concrete).

From the builder's perspective, cost may also be an issue. Whether to have wet finished plastered walls (*in-situ* plaster) or whether to dry line the walls with plasterboard may have a significant effect on speed and hence cost. Consideration of which finishes to apply, although limited in choice range, may be viewed through different eyes: for example, the home owner or the builder.

When we examine buildings other than houses, a number of other criteria emerge. We may want to know, for example, the lifespan of the finish, its maintenance requirements, its acoustic properties and its non-slip qualities. We may view the total cost for finishes (particularly the heavily burdened floor finishes) by examining the initial purchase and laying cost, together with the cleaning cost, maintenance cost, replacement cost, disruption and so on. In this way, performance is considered over the life of the building.

12.2 | Wall finishes

Introduction

- After studying this section you should be able to identify the nature of *in-situ* wall plasters used in residential property.
- You should understand the use of plasterboard dry lining finishes as an alternative to wet wall finishes.

Wall finish options

Wet plaster is the traditional way of finishing walls for houses. Over the years, plaster has been formed from a variety of materials: sand/lime, fibrous hair reinforced and so on. The modern and most widely used plaster today is gypsum based. BS 1191 outlines the specification for these plaster types, which are also often referred to as 'calcium sulphate plasters'.

In its unretarded state, this form of plaster tends to be of little use apart from in situations requiring the material to set in minutes. Until recently, these types were used in hospitals for the setting of broken bones. Once a retarder is added to the mix to slow the setting time, the plaster is suitable for applying to walls and ceilings.

One of the processes undertaken to create the range of gypsum plasters covered by BS 1191: Part 2 is the application of heat. The chemical ingredients of gypsum or calcium sulphate contain two molecules of water: hence $2H_2O$. If this is heated until three-quarters of the water evaporates, half a molecule of water is left, and this form of the plaster which is extensively used is now termed 'hemi-hydrate gypsum plaster'. Of course we have already mentioned the retarder, so retarded hemi-hydrate tends to be the form that the plaster takes.

The British Standard also sets out the number, type and thickness of coats of plaster recommended for different surfaces. As a result, a single coat or *skim coat*

Figure 12.1
The external angle bead.

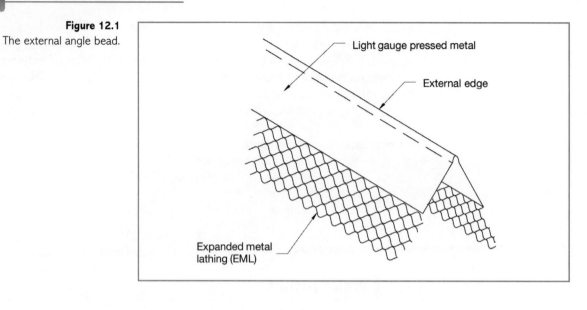

Light gauge pressed metal

External edge

Expanded metal
lathing (EML)

Figure 12.2
The external angle bead
in position.

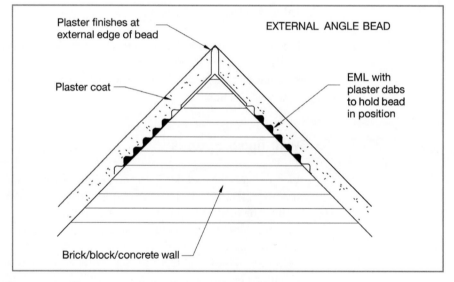

Plaster finishes at
external edge of bead

EXTERNAL ANGLE BEAD

Plaster coat

EML with
plaster dabs
to hold bead
in position

Brick/block/concrete wall

Figure 12.3
Covering the
plaster/lining junction
with an architrave.

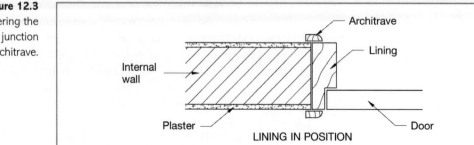

Architrave

Lining

Internal
wall

Plaster

Door

LINING IN POSITION

tends to be recommended for application to plasterboard, a two-coat application to blockwork and perhaps a three-coat application to concrete (depending on its density).

The normal thickness of two-coat work to blockwork is 13 mm overall, and this is the most widely used specification in housing owing to the popularity of block-work.

It should be remembered also that the two coats tend to be in different plasters – undercoat plaster typically 10 mm thick, and a harder finish plaster 3 mm thick.

Plaster accessories

Metal accessories are available to reinforce certain locations on plaster that may be vulnerable to damage or where shrinkage cracks are anticipated. These include the external angle bead and the stop bead.

Both the external angle bead and the stop bead are formed of light gauge pressed metal. The solid metal portion of the bead provides the edge required, while the remainder is slotted and stretched to form expanded metal lathing (EML). Figure 12.1 is typical.

The expanded metal portion of the bead provides the means of fixing this bead to the wall using plaster dabs. Having located the bead on the corner of the wall or partition, the plaster undercoats and finish coat are worked up to the edge, as shown in Figure 12.2.

When a timber lining is used as the frame, the usual way to finish the wall plaster and frame junction is to cover the meeting with a timber architrave. The reason for this is that the plaster will always shrink back from the timber lining and leave an unsightly crack, which needs to be covered (Figure 12.3).

Occasionally, it may be preferable to extend the door lining through the wall beyond the plaster surface, and this tends to make a feature of the lining. When this detail is used the possibility of using an architrave is discounted, but the crack which is expected between the lining and the plaster can be controlled by the use of a stop bead (Figure 12.4).

> **REVIEW TASKS**
> - What is the basic chemical ingredient of gypsum plaster?
> - How many layers and what thickness would be appropriate when applying plaster to a blockwork surface?
> - Visit the companion website at www.palgrave.com/engineering/riley1 to view sample outline answers to the review tasks.

Dry lining walls with plasterboard

Instead of using *in-situ* (wet) plaster to finish internal walls to a property, we can dry line the walls with plasterboard sheets. The plasterboard may be fixed to the wall in

PART 3

Figure 12.4
The use of a metal stop
bead.

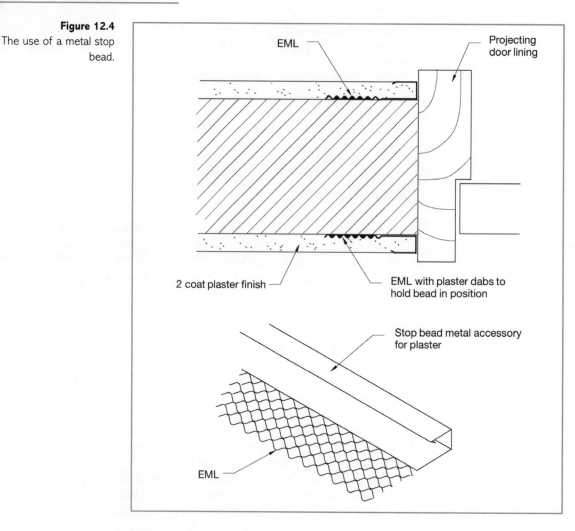

Figure 12.4
The use of a metal stop
bead.

a variety of ways, but whichever method is chosen this technique may be considered to have a number of advantages over wet plaster:

■ It creates a small void between the back of the plasterboard and the backing wall, which can be useful for threading cables for the electrical installation

■ The air void is a good insulator and this may assist the thermal insulating qualities of the wall

■ The thermal insulation qualities of the wall may be considerably assisted also by the use of thermal plasterboard, which can be supplied with insulation (for example, polystyrene) bonded to the back (for example, Gyproc Thermal Board, British Gypsum)

■ Dry plasterboard means quicker completion of the wall finish with no real delay for the drying time which would be associated with wet plaster.

The three methods of fixing the plasterboard sheets to the wall include:

■ Fixing with adhesive dabs

Figure 12.5
Dry lining – fixing plasterboard with adhesive.

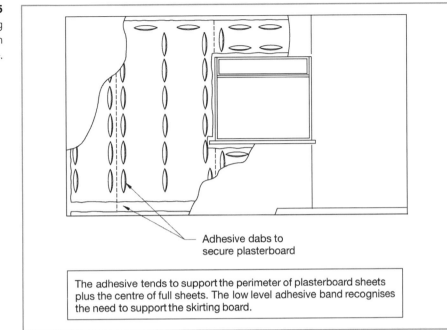

Adhesive dabs to secure plasterboard

> The adhesive tends to support the perimeter of plasterboard sheets plus the centre of full sheets. The low level adhesive band recognises the need to support the skirting board.

- Fixing onto a sawn timber batten frame (wedged vertical from the surface of the wall and sometimes preferred for uneven existing walls)
- Fixing on to a proprietary metal channel (metal furring system, British Gypsum).

Fixing by plaster dabs is illustrated in Figure 12.5. As shown, the adhesive dabs tend to be located at the perimeter of the plasterboard sheets.

Figures 12.6, 12.7 and 12.8 show the detailing where dry lining meets a door lining, a window opening and a typical external corner.

Figure 12.6
Dry lining junction with an internal door lining.

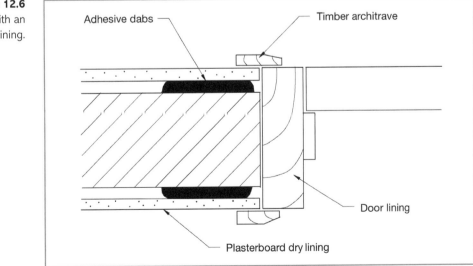

Adhesive dabs

Timber architrave

Door lining

Plasterboard dry lining

PART 3

Figure 12.7
Dry lining meeting with a
window frame.

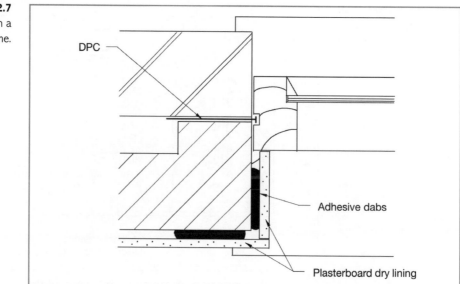

DPC

Adhesive dabs

Plasterboard dry lining

An alternative to fixing plasterboard with adhesive dabs is to use the metal furring system. This uses light gauge metal channels which are fixed to the wall in a proprietary adhesive and set vertical using a spirit level. The centres of the channels coincide with the width of the plasterboard sheets, and fixing plasterboard is by self-tapping screws through the plasterboard and through the channel (Figure 12.9).

With dry lining techniques, the tendency is not to cover them with a single skim coat of plaster but instead to use better quality plasterboard sheets which have an ivory card face to the outside for direct decoration. With these plasterboards the fixture screwheads are filled and the face of the board rubbed with a proprietary slurry (sponge applied) to present an even surface on which paint can be directly applied.

Figure 12.8
Dry lining to an external
corner.

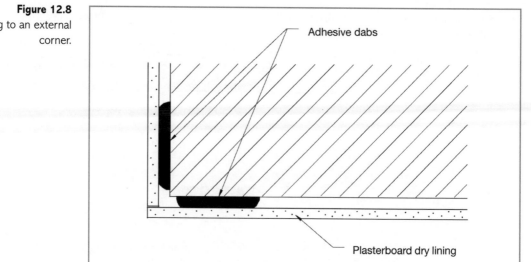

Adhesive dabs

Plasterboard dry lining

Figure 12.9
Dry lining – the metal
channel system.

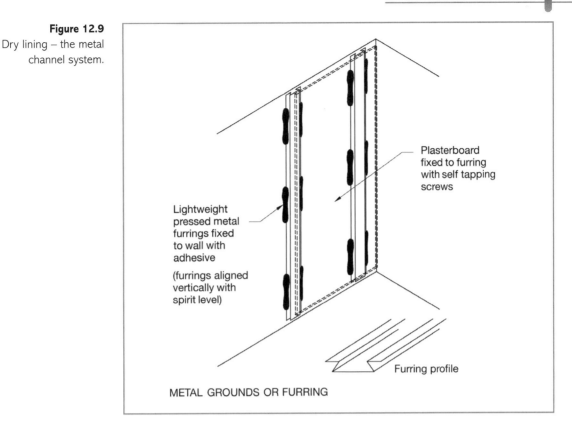

Plasterboard
fixed to furring
with self tapping
screws

Lightweight
pressed metal
furrings fixed
to wall with
adhesive

(furrings aligned
vertically with
spirit level)

Furring profile

METAL GROUNDS OR FURRING

Wall tiling

Ceramic wall tiles are still popular for certain rooms of the house, such as the bath-room, kitchen or WC, where tiling may be full floor-to-ceiling height, or to only part of the wall in the form of a dado (Figure 12.10).

As an alternative to large areas of tiles as illustrated, tile splashbacks are also often provided behind sinks and hand basins.

BS 8000: Part 11 covers the fixing of these tiles to walls and floors, while BS 6431 refers to tile quality.

Today, tile adhesive is often also capable of use as grout between the tiles. While the tiles themselves are impervious, it is important to recognise the need to make the joints between them watertight with grout.

Splashbacks and glass/steel alternatives

Tastes in the internal finishes to dwellings change constantly and the solutions chosen for dealing with wall tiling, splashbacks and so on have evolved greatly in recent times. It is now common to see the utilisation of metal, glass and plastic splashback details as well as traditional tiling.

Figure 12.10
Wall tiled dado to a
bathroom or kitchen.

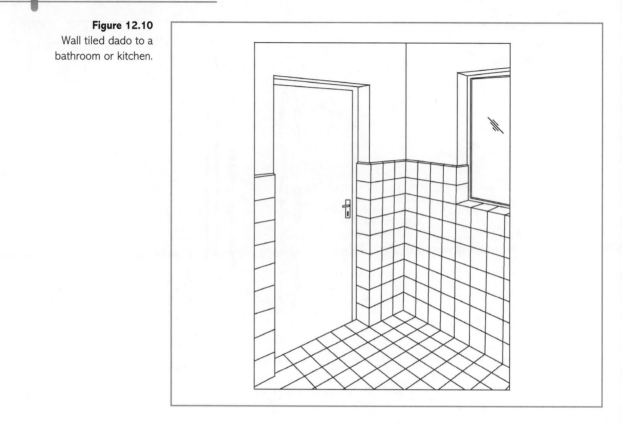

12.3 | Ceilings and ceiling finishes

Introduction

■ After studying this section you should appreciate the forms of finish to ceilings.

Ceiling finish materials

Before the availability of plasterboard, ceilings to houses tended to be formed using the lath and plaster technique. Thin strips of sawn timber (not planed) were nailed to the underside of the joists with gaps left between each timber strip. The plasterer could then apply the plaster to the ceiling and force some of the material between the laths where it would achieve a key by swelling, as shown in Figure 12.11.

Modern ceilings, in comparison, tend to be formed in plasterboard sheets which are clout nailed (galvanised mild steel nails with large heads) to the joists. Joists are usually set at 400 mm centres for floors and for ceiling joists of roof solutions, as this suits the plasterboard sheets, which are produced in multiples of 400 mm. The largest sheets are 2,400 mm × 1,200 mm.

Figure 12.11
Ceiling options.

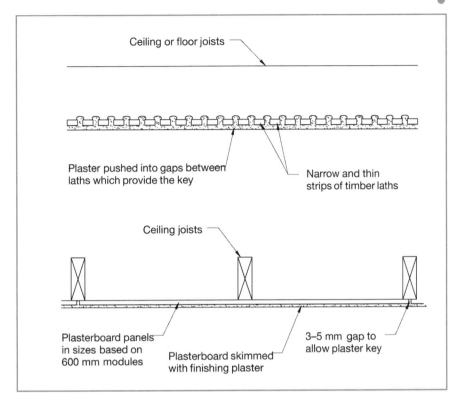

Ceiling or floor joists

Plaster pushed into gaps between laths which provide the key

Narrow and thin strips of timber laths

Ceiling joists

Plasterboard panels in sizes based on 600 mm modules

Plasterboard skimmed with finishing plaster

3–5 mm gap to allow plaster key

Setting the joists at 400 mm will ensure that there will always be timber against which the edge of the plasterboard can be nailed. Alternatively, 600 mm joist centres would suffice provided this satisfies the structural needs of the span in question.

A skim coat of plaster (less than 5 mm thick) is the normal way of finishing the plasterboard ceiling. Before the application of the plaster, the joints between the plasterboard sheets are covered with a *scrim* reinforcement to resist cracking of the ceiling plaster along the lines of the plasterboard sheet edges. This scrim used to be in jute cloth, but the modern equivalent is plastic-based self-adhesive netting material applied across the joints from narrow rolls. Any deflection to the floor above the ceiling would otherwise tend to cause fracture of the plaster at the position of weakness: at the plasterboard sheet edge.

Plasterboard is covered by BS 1230. It is simply gypsum plaster sandwiched between two layers of card. The normal card is slightly rough to the touch and this helps the key to the wet plaster skim coat. Alternatively, the plasterboard can be obtained with a better quality ivory card face to one side to allow the board to be directly decorated rather than plastered.

The thickness of plasterboard sheets depends on the type of plasterboard in question but may be 9.5 mm, 12.5 mm or 19 mm thick. British Gypsum is the main plasterboard supplier and this company uses the name Gyproc as its trade name for plasterboard. Gyproc Wallboard is just one variety. The thickest sheet without attached insulation is Gyproc Plank, at 19 mm.

REVIEW TASKS

■ What thicknesses are available in plasterboard sheeting?

■ What sort of metal bead would you use to reinforce external plaster corners on a wall?

■ What is the purpose of scrim reinforcement?

■ Visit the companion website at www.palgrave.com/engineering/riley1 to view sample outline answers to the review tasks.

12.4 | Floor finishes

Introduction

■ After studying this section you should appreciate some of the timber finishes that may be applied to floors.

Overview

As stated in the Overview to section 12.1, the scope for floor finishes in the home is somewhat limited. Many properties are constructed and sold without any floor finishes at all, just the floor surfaces as constructed: either timber boards, chipboard sheets, power-floated concrete or screeded concrete. These of course will tend to be subsequently carpeted or similar, to the requirements of the owner.

This brief section intends to focus on finishes that may be applied prior to the sale of the unit.

Timber finishes

There are really two timber solutions that might be regarded as traditional finishes to floors in dwellings: wood block flooring and hardwood strip flooring.

Whichever of these is used it should be remembered that timber is hygroscopic and as such will absorb moisture. This moisture may be from the air of the room in question or it may be from physical contact with moisture. In the latter case, we need to ensure that suitable damp-proof course material occurs between the finish and the moisture source.

When timber absorbs moisture, its moisture content rises and expansion results. There is a need therefore to recognise the fact that timber floor finishes will move in use, and provision for movement must be incorporated into the laying procedure.

When hardwood strip flooring is used, a degree of restraint may be gained by the nailing that often takes place in securing the strips to the joists of the floor structure (Figure 12.12).

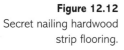

Figure 12.12

Secret nailing hardwood strip flooring.

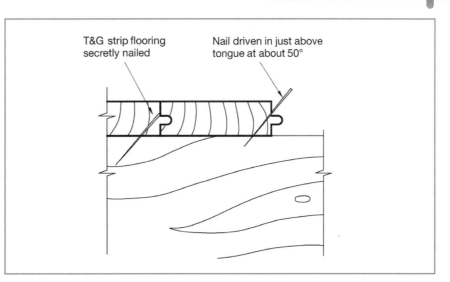

The extent of the gap left around the perimeter of strip flooring will depend to a degree on the type of hardwood that is used. A significant difference between the dimensional changes caused by ventilation in moisture content exists between different hardwoods.

If the perimeter expansion gap is not to be covered by a skirting board, compressible fill may be used in the gap.

REVIEW TASKS

- Describe the secret nailing fixture method as applied to boarded floor finishes.

- What are the *two* popular patterns for laying wood block flooring?

- What distinguishes parquet flooring from wood block flooring?

- Visit the companion website at www.palgrave.com/engineering/riley1 to view sample outline answers to the review tasks.

13

Overview of alternative sustainable construction methods

AIMS

After studying this chapter you should be able to:

- Recognise the materials and technologies that are available which promote sustainable construction and green buildings
- Discuss the alternative construction methods that are being developed for dwellings
- Appreciate the benefits and features of a range of alternative technologies used for constructing houses and other buildings

This chapter contains the following sectons:

13.1   Sustainable construction
13.2   'Natural green' technologies and ICF

INFO POINT

- Anderson, J., Shiers, D. and Steele, K.(2009) *The Green Guide to Specification*, 4th edn, Wiley-Blackwell, Oxford [ISBN 978–1–848–06071–5]
- Bevan, R. and Woolley, T. (2008) *Hemp Lime Construction, A guide to building with hemp lime composites*, IHS BRE Press, Garston, Watford [ISBN 978–1–848–06033–3]
- Carroll, B. and Turpin, T. (2002) *Environmental Impact Assessment Handbook*, Thomas Telford Publishing [ISBN 0–727–72781–8]
- CIBSE (2007) *CIBSE Knowledge Series: Green Roofs*, The Chartered Institution of Building Services Engineers, London, September [ISBN 978–1–903–28787–3]

- Communities and Local Government (2011) *Zero Carbon Non-domestic Buildings, Phase 3 final report*, DCLG Publications, July, Crown Copyright
- Construction Products Association (2007) *Delivering Sustainability: The Contribution of Construction Products*
- Hartman, H. (2007) 'A Subtle Green', *Architects Journal*, AJ 08.11.07, pp. 38–40
- HM Government Low Carbon Construction Innovation and Growth Team (2010) *Final Report*, Department for Business, Innovation and Skills, Crown Copyright
- Kibert, C. (2005) *Sustainable Construction: Green Building Design and Delivery*, Wiley
- Woolley, T. and Kimmins, S. (2000) *Green Building Handbook, Volume 2*, Spon [ISBN 0–419–25380–7]

13.1 | Sustainable construction

Introduction

This chapter aims to give a brief overview of some of the technologies that are available to undertake sustainable construction of dwellings. The explanation of some of these available techniques and technologies is not intended to be exhaustive. However, it is hoped that the reader will be able to develop an insight into the use and benefits of these alternative technologies, which are increasing in popularity. One of the main aims of this chapter is to raise your awareness of what green buildings actually are and to give you ideas that could be utilised in the future as alternative sustainable approaches to the construction of houses. Hopefully you will find this chapter interesting enough to make you want to investigate the topic of green buildings further.

Overview

At the outset of this chapter it is essential to appreciate that the entire construction industry is becoming more sustainable. Changes driven by legislation, user awareness and energy conservation requirements are combining to increase the degree to which all house building is increasingly 'sustainable'. As such, various technologies that have been discussed throughout this book may be described as sustainable. In particular it must be stressed that MMC, timber framed and steel framed designs are deemed to be sustainable construction approaches. However, they are what might be considered 'mainstream' construction technology. This chapter attempts to present some of the alternative technologies that are still emerging and, as yet, have not been utilised for high volume construction solutions.

Many of these technologies are well established and may actually be considered as historic methods of building. They often reflect small carbon footprints and may utilise readily available, locally sourced materials. Alternatively, they may adopt industrialised approaches to resource minimisation, controlled materials use and low-energy construction and use of buildings. Several of the elements that we have

PART 3

previously described have touched on sustainable options and it is not intended to repeat that material here. However, it is worthwhile considering some elements and construction methods specifically from the perspective of sustainability rather than just general performance requirements. Hence, there may be some element of overlap although the specific focus upon the 'green' aspects of their design and use is the primary intention within this chapter.

Foundations

The size and design solution of foundations are generally driven by the scale, predicted loads and structural form of a building, together with its location and the ground conditions. The main types of foundation used for dwellings have been described in detail in Chapter 4 and they include strip foundations, trench fill, rafts, pads and piles. When considering the design of foundations from a sustainability perspective there are several principles that need to be adhered to. Careful scrutiny of the ground conditions should be undertaken to avoid overdesign, which could result in more concrete and steel being used than is necessary. Alternatives that utilise less material should be considered but also designs that result in less excavation and therefore less tipping of soil to landfill. Careful consideration of the materials to be used can also aid sustainability, and the use of recycled materials such as pulverised fuel ash will reduce the amount of cement required.

By more careful consideration of the materials used to construct the superstructure, the weight of a building can be significantly reduced, so diminishing the size of the foundations needed. This is one of the advantages of MMC, timber and steel framed house construction.

When using mass (wet) concrete in foundations, it is very common to have a trench that is bigger than required and therefore more concrete is needed. This is partly because the buckets on mechanical excavators come in a standard range of sizes and the width of the trench may be a result of the size of the bucket rather than the minimum required foundation width. In low-rise housing it is common for very modest foundations to be required, however the trench width of 600 mm became almost a default width as a result of excavator bucket width. Using a proprietary system that is produced in a factory removes this problem. Figure 13.1 shows such a system where piles have been driven into the ground and a precast ground beam spans between to support the external walls. This system uses minimum materials, is relatively lightweight and requires no excavation if the existing ground is at the correct reduced level to start with. Additionally, the materials used in this system can all be easily recycled if or when the building is demolished.

Structural frames

In domestic construction, virtually all house builders use timber frame for houses (Figure 13.2). The benefits of timber frame for domestic construction from a sustainability perspective are well known. The use of lightweight steel framed construction

Figure 13.1
Proprietary foundation
system.

Figure 13.2
Timber framed house
construction.

is also growing in popularity and the potential for recycling of steel has increased the acceptance of this as a sustainable construction method. The light weight of the super-structure, the speed of construction, with minimal materials wastage and the potential for high insulation and air-tightness all combine to make these approaches among the most sustainable of the mainstream construction methods used for housing.

13.2 'Natural green' technologies and ICF

Clay wall construction

Clay wall construction is not new and is used extensively around the world as a vernacular form of construction. It is an extremely green method of construction

PART 3

Figure 13.3
Foundation/slab/wall
details for clay wall
construction.

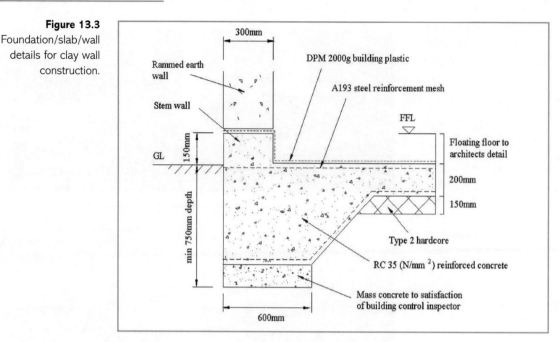

and there are numerous examples of its use in the UK and in mainland Europe, especially in the south-east of England which sits on a large clay cap, which means that the materials required are naturally and abundantly available.

A foundation/slab is formed as with traditional construction but a kicker is formed along the lines of the proposed wall. Wall formwork is then erected and clay of layers of 100–150 mm is placed in the formwork. This is then mechanically rammed to ensure it is compact. The process is then repeated until the wall is the required height. The wall is cast in panels and expansion joints are incorporated between the panels. Once all the panels are complete, the wall is then topped with a concrete ring beam onto which the roof structure is fixed. Figure 13.3 shows the foundation to wall configuration used in clay wall construction.

Figure 13.4 shows sliding formwork being used to help form the walls and Figure 13.5 shows typical external wall/roof details for a clay wall building with a traditional pitch roof.

Straw bale wall construction

Straw bale wall construction is a system that is becoming more common in the UK. For the building shown in Figure 13.6, a timber frame has been constructed on a traditional slab and straw bales used to fill in between the frame elements. The straw has then been rendered over to form the finished wall. Straw is an abundant and highly renewable material and therefore this type of construction is extremely green. Figure 13.6 shows the timber frame, the straw contained in the wall and the first coat of render. The inclusion of this detail has become traditional in straw bale construction and it is known as an 'honesty window'.

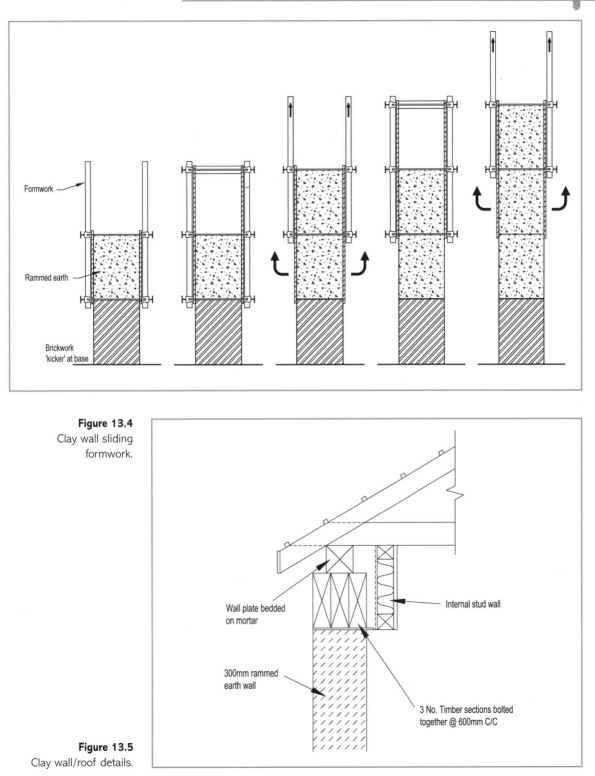

Formwork

Rammed earth

Brickwork
'kicker' at base

Figure 13.4
Clay wall sliding
formwork.

Wall plate bedded
on mortar

Internal stud wall

300mm rammed
earth wall

3 No. Timber sections bolted
together @ 600mm C/C

Figure 13.5
Clay wall/roof details.

PART 3

Alternatively, the straw bales may be outside the timber frame rather than between the elements and they may be clad with timber or other materials rather than rendered. They can also be used with alternative foundation and flooring systems which are more sustainable that traditional wet concrete systems.

Hemp lime/hempcrete

Hemp lime is a 'novel' construction material. The composite material combines fast-growing renewable and carbon-acquiring plant-based aggregates (hemp shiv) with a lime-based binder to form a lightweight material that is suited to various construction applications, including solid walls, roof insulation and under-floor insulation, and as part of timber-framed building. It also offers good thermal and acoustic performance and the ability to regulate internal relative humidity through hygroscopic material behaviour, contributing to healthier building spaces and providing effective thermal mass.

The hemp shiv is mixed with a lime-based binder which binds the hemp aggregates together, giving structural strength and stiffness. The lime also protects the shiv from biological decay. Interestingly, this material can be used to manufacture solid non-loadbearing panels, which can be installed as the cladding to a timber (or other) framed building. A number of studies have shown that hemp lime has good thermal and acoustic insulation properties. The lightweight hemp lime absorbs sound, dampening transmission through walls and other elements.

Hempcrete is a mixture of hemp hurds (woody core) and a hydroscopic (breathing) binder, usually lime based which produces an insulating and heat-storing material which ideally is used to build the walls, floor and infill the roof structure to provide a very-low-energy house which is very healthy to live in. Figure 13.7 shows a hemp house under construction. As you can see, this will look just like a traditional house. Figure 13.8 shows the finish to the external walls when sprayed hempcrete has been applied.

Cladding using hemp lime/hempcrete can be constructed using techniques similar to those for clay walls, using formwork, or it can be formed into blocks and the

Figure 13.7
Hemp house under
construction.

Figure 13.8
External walls coated with
sprayed hempcrete.

PART 3

cladding constructed by means of techniques similar to those utilised in traditional blockwork. The hemp lime can also be used for roofing when formed into sheets.

The main advantages of hemp lime are:

- It is a means of achieving energy efficiency due to its insulating properties
- It is a breathable material which can help to create healthy buildings
- It is made from renewable sources and creates little environmental damage
- There is no pollution created at end-of-life disposal
- Hemp is a crop-based material, which helps farmers, and is a good use of land
- It offers the possibility of acquiring carbon into building fabric as opposed to emitting it
- Hemp lime has the ability to make an impact on the future of sustainable building by reversing the damaging effects of greenhouse gases. It is claimed that hemp lime can lock up approximately 110 kg of carbon dioxide per cubic metre of wall.

Figure 13.9
Traditional wall with hemp
lime inner leaf.

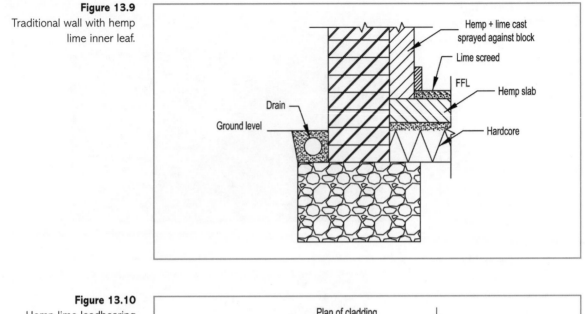

Figure 13.9
Traditional wall with hemp
lime inner leaf.

Figure 13.10
Hemp lime loadbearing
wall with rain screen.

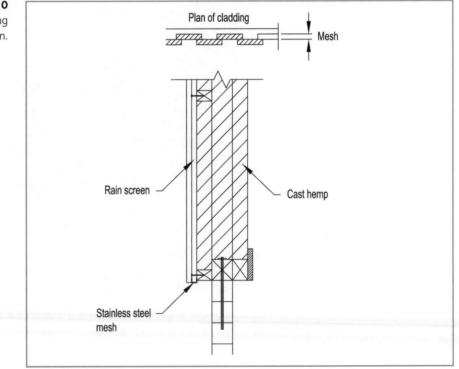

Figure 13.10
Hemp lime loadbearing
wall with rain screen.

Figure 13.9 shows a traditional wall with an inner leaf of hemp lime cast onto it and Figure 13.10 shows the main loadbearing wall constructed using hemp lime with a rain screen attached to it to prevent damage. These are two potential situations where the use of hemp lime would be recognised.

Figure 13.11
'Polysteel' insulated concrete formwork system.

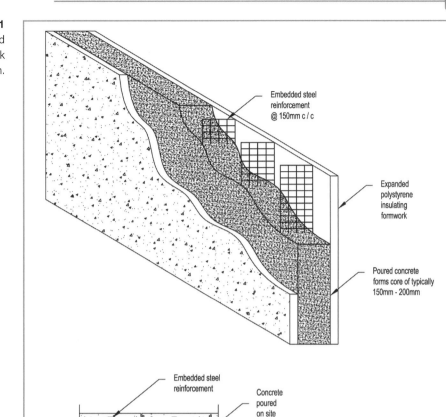

Embedded steel
reinforcement
@ 150mm c / c

Expanded
polystyrene
insulating
formwork

Poured concrete
forms core of typically
150mm - 200mm

Embedded steel
reinforcement

Concrete
poured
on site

Insulated EPC form
with interlocking
adjacent sections

Insulated concrete formwork (ICF)

Insulated concrete forms (ICFs) are intended to provide a highly energy-efficient method for constructing the external walls of dwellings and other small to medium scale buildings. The system utilises blocks made from flame-retardant expanded polystyrene to create permanent insulating formwork into which mass concrete is poured to create a robust and highly insulative external enclosure. The formwork blocks are manufactured in a range of standard modular sizes, fitting together using a tongue and groove design. The blocks interlock to form the wall profiles into which mass concrete is poured. The concrete is reinforced using steel cage sections that link to the inner and outer sections of the polystyrene formers. Once the concrete has cured, these walls are designed to have high strength, typically several times stronger than for walls built using traditional methods in high-density concrete blocks or bricks. There is no need for additional insulation as the insulation from the former and the concrete is sufficient to better the insulation levels required by

PART 3

Figure 13.12
Construction sequence
for insulated concrete
formwork (ICF).

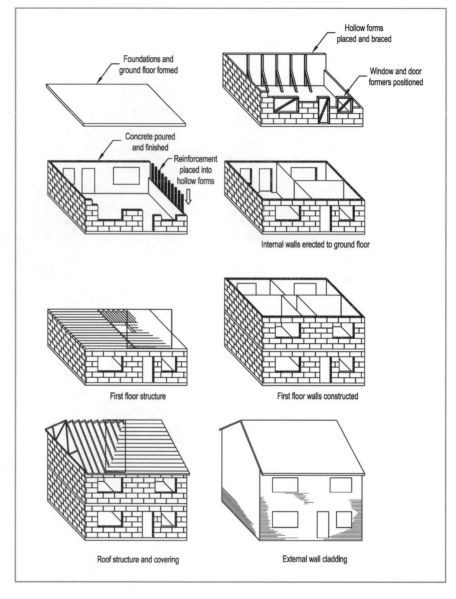

Building Regulations. Figure 13.11 illustrates the construction form of walls built using the 'Polysteel' ICF system.

The main benefits of this type of construction are as follows:

- Fast construction (it is possible to build up to 5 m² of wall area per man hour)
- Low levels of materials wastage
- Easy on-site handling of lightweight units
- Creation of high-mass, high-strength walls, making it a good option for basement and retaining walls
- Enhanced air-tightness due to the homogeneous construction form utilised

- High sound insulation
- High fire resistance.

Figure 13.12 illustrates the typical sequence of operations involved in the construction of properties using this method. The nature of this technique is such that it affords great flexibility in design and is often used for individual dwellings. Figures 13.13 to 13.16 illustrate some of the key features of the technology that is used in ICF construction.

Figure 13.13
ICF formers placed and braced, awaiting concrete placement within hollow formers.

Figure 13.14
Pumped concrete placed within ICF formers.

Figure 13.15
Walls formed with high-strength concrete placed within formers.

PART 3

Figure 13.16
Dwelling nearing
completion.

Green roofs

There is a growing recognition of the benefits of increased landscaping and natural features in the design and construction of buildings. Such features are known to have tangible benefits in terms of overall environmental quality and sustainability.

One way of increasing the amount of landscaping without increasing the footprint of a development is by the inclusion of what is known as a 'green roof'. Not only do roofs of this type increase the amount of landscaped area in a development, but they can also be developed as relaxation areas for occupants. They also lead to significant cooling in buildings and are becoming increasingly popular in warm climates.

These green roofs can be formed using traditional turf but are usually formed using 'sedum' as the top layer. Sedum is a genus of plants which have water-storing leaves and are therefore ideal for green roof construction. Water that falls during periods of high rainfall can be stored by these plants to irrigate the roof during dry periods.

The most common types of green roof are:

■ Extensive, which means they cover the whole roof and the plants are delivered to the site from out of the area, and this ensures that they will have a consistent appearance, as shown in Figure 13.17.
■ Extensive but biodiverse, which means that the plants that form the top surface will not be planted at the time of construction but will seed naturally. This may lead to an inconsistent appearance but will have a high level of biodiversity.
■ Intensive, which are sometimes called roof gardens. The plants are used to form a 'park' on the roof, which can be used as an amenity as well as providing all the other benefits of turf roofs.

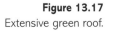

Figure 13.17
Extensive green roof.

Figure 13.18 shows the typical elements of a green roof build-up. A slab is constructed and then a layer of insulation applied. As there will be significant moisture held in the top layer, a waterproof membrane is required. To prevent the plant roots damaging the waterproof membrane, a root barrier layer is required. In order to prevent the top layer from drying out, a water-retention layer needs to be included that complements the water-retention properties of sedum plants and stores water during wet periods for use when the weather is drier. Lightweight soil is used to reduce the load on the structure and shallow root system plants are used to form the top layer. These plants need to be durable and easily regenerated to ease the maintenance requirements of the roof. Figure 13.17 showed a single-storey building with a turf roof. However, this type of roof is increasingly being used in multi-storey buildings. Figure 13.19 shows an alternative method of forming green roofs using pre-planted sedum trays that are laid onto the roof membrane.

Figure 13.18
Eco green roof build-up.

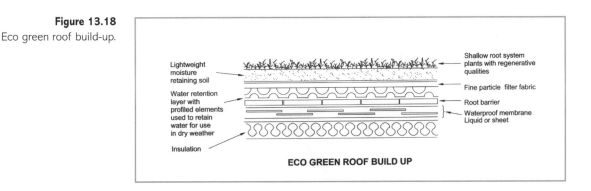

Lightweight moisture retaining soil

Water retention layer with profiled elements used to retain water for use in dry weather

Insulation

Shallow root system plants with regenerative qualities

Fine particle filter fabric

Root barrier

Waterproof membrane Liquid or sheet

ECO GREEN ROOF BUILD UP

PART 3

Figure 13.19
Sedum trays used to form a 'green roof' (courtesy of Denovo Design).

The case study that is described here is based on the construction of dwellings for social housing and was undertaken by Liverpool Architects Denovo Design on behalf of Halton Housing. It was designed to achieve Level 4 of the *Code for Sustainable Homes*. The scheme utilises a traditional construction form supplemented by a range of environmental features that are aimed at delivering a sustainable end-product. The main sustainability-related design interventions are summarised below.

Energy demand within the dwellings is reduced by:

■ Incorporating high levels of insulation
■ Utilising timber frame construction to provide a weather-tight and airtightness-tested envelope
■ Installing high-efficiency gas-fired boilers to provide heating and hot water
■ Providing low-cost, continuous whole-house ventilation, which moderates temperature and assists in preventing condensation
■ Using photovoltaic panels to generate electricity
■ Adopting sun tunnels and relatively large window areas to make best use of natural light.

Demand for water was reduced by installing:

■ Low-water-content radiators
■ Low-flow showers and spray taps, fitted with flow regulators
■ Low-water dual-flush WCs
■ Water butts to collect and allow reuse of rainwater.

Use of sustainable materials was maximised by using, for example:

■ Recyclable galvanised mild steel for railings, rainwater goods and door numbers
■ Water-based interior paints and water-based timber preservatives
■ High-performance softwood from proven FSC accredited sources for all windows and doors.

Figures 13.20 to 13.24 illustrate various aspects of this case study.

Figure 13.20
Code 4 sustainable home.

Figure 13.21
Timber frame construction clad with masonry.

Figure 13.22
Photovoltaics and sun tunnels.

Figure 13.23
Photovoltaic panels on roof and sun tunnel to utilise natural light.

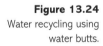

Figure 13.24
Water recycling using water butts.

PART 3

REFLECTIVE SUMMARY

■ A range of alternative construction options is available for dwellings that might be considered sustainable. Many of these offer realistic solutions for small-scale developments rather than mainstream construction projects.

■ When deciding what materials and systems are to be used to construct a building with green elements, a whole-building approach needs to be taken rather than considering elements in isolation.

■ The construction of the structure of a building can have as much an impact in creating a sustainable building as the services installation. Embodied energy must always be considered when selecting design alternatives.

■ Green roof, clay wall, straw bale wall and hemp wall construction are examples of technologies that are not only labelled as 'green', but as 'natural green' because they use natural materials and systems in their construction.

■ Systems that adopt high-tech solutions and industrialised construction methods may be considered sustainable because of their ability to create low-energy buildings and reduce waste during construction.

REVIEW TASKS

■ If we know that there are natural green technologies available for the construction of walls and roofs, why are they not used more widely?

■ Consider the recent construction of houses in the area where you live. Can you identify any features that are clearly intended to promote sustainability?

■ Compare and contrast the construction of dwellings that were built in the 1900s, the 1930s and the 1970s with those featured in the case study. What are the main differences between these various forms that relate to sustainability and our changing approach to environmental issues?

■ Visit the companion website at www.palgrave.com/engineering/riley1 to view sample outline answers to the review tasks.

Index